RSPB SEABIRDS

RSPB SEABIRDS

Marianne Taylor
Photographs by David Tipling

BLOOMSBURY
LONDON • NEW DELHI • NEW YORK • SYDNEY

The RSPB is the country's largest nature conservation charity, inspiring everyone to give nature a home so that birds and wildlife can thrive again.

By buying this book you are helping to fund The RSPB's conservation work.

If you would like to know more about the RSPB, visit the website at www.rspb.org.uk, write to The RSPB, The Lodge, Sandy, Bedfordshire, SG19 2DL, or call 01767 680551.

First published in 2014

Bloomsbury Publishing Plc
50 Bedford Square
London
WC1B 3DP

www.bloomsbury.com

Bloomsbury is a trademark of Bloomsbury Publishing Plc

Bloomsbury Publishing, London, New Delhi, New York and Sydney

A CIP catalogue record for this book is available from the British Library

UK ISBN (hardback) 978-1-4729-0901-5
UK ISBN (ePUB) 978-1-4729-1116-2

10 9 8 7 6 5 4 3 2 1

Commissioning editor: Julie Bailey
Design by Julie Dando, Fluke Art

Printed in China by C&C Offset Printing Co., Ltd.

Page 1 (Eider) and page 2 (Kittiwakes) photographs both by David Tipling

Contents

Introduction

Imagine sitting in a small boat on calm water. Masses of Gannets wheel overhead, huge and angelic, treading air with hardly a flicker of their white, black-tipped wings. As you watch one, it tilts downwards, folds in those long wings and, drawing them right back to turn itself into a spear, dives in a headlong, high-speed vertical plunge, punching a neat hole in the water and throwing up a plume of white spray. You watch its vapour trail of bubbles twist and turn as it chases after a fish. All around you others are doing the same, almost close enough but much too fast to touch. Later, you wait on a grassy island slope through the long summer evening as rafts of Manx Shearwaters gather offshore, the warm evening light flushing their white chests rosy. As darkness falls they head for land; suddenly they are everywhere, blundering past you to their burrows, shouting to their chicks underground, the cacophony of their bizarre voices growing as the night takes over, until you are alone in an alien soundscape.

The people of our island nation have long drawn inspiration from the sea. Its richness, power and mystery infuse our identity, and many of us also have a special respect and affection for the birds that make their living as seafarers. Like the sea itself, seabirds seem to have a special wildness and freedom compared with even the most elusive land birds, and yet are among the easiest of all our birds to observe at close quarters, offering wildlife-watching experiences of incredible intensity.

Compared with many countries in mainland Europe, the UK and Ireland have a rather sparse, impoverished wildlife population. The great mammalian predators of our forests are long gone, and many of the mainland's plants and small animals simply never reached us. Due to our dense human population true wilderness is hard to find. However, turn away from the land, look to the sea, and suddenly we are punching well above our weight in terms of wildness.

Great Britain has about 18,000km of coastline. Add Ireland's coastline and that is another nearly 2,800km. The hundreds of smaller islands all around our coastline almost double this total. Along that great length of coastline there are beaches of boulder, shell and stone; sheer chalk and granite cliffs; softly crumbling slopes of sandstone and mud; dunes anchored by marram grass into towering crested ridges; estuarine mudflats and saltmarshes, and seaside towns of every character. All of these provide habitats for wildlife, as of course does the open sea itself.

Our seabirds, perhaps more than any other animals, have a foot in both worlds, needing dry land to nest and rear their young, but relying on the sea for their foraging. Every detail of their anatomy, from the colours of their plumage to the structure of their bones, speaks of adaptation to a life in intimate association with the open sea, and their behaviour is no less specialised. Most seabirds are intelligent and long lived, many form pair bonds that can endure for decades, while two of our seabirds – the Arctic Tern, which breeds here, and the Sooty Shearwater, which is a regular visitor to British coasts – are extreme long-distance travellers, each regularly flying 65,000km or more in a year.

The British Isles are of global importance for their breeding seabirds. It includes more than half the world's entire breeding populations of three species – the Manx Shearwater, Gannet and Great Skua. Of the remaining 22 seabird species that regularly breed around our coasts, 18 in Britain and eight in Ireland have internationally important populations. Outside the breeding season, the seas around the British Isles support huge numbers of other seabird species that breed inland or on coasts elsewhere.

A WILDERNESS OF WATER

The Atlantic Ocean surrounds us on all sides and extends across more than 20 per cent of the world's surface. Marine life around the British coastline is both plentiful and diverse. The North Sea supports one of the world's richest and most important fisheries, and the mackerel, cod, whiting and plaice that the boats bring in each day are a tiny part of a huge ecological web, representing thousands of genera across the full sweep of biological life. Seabirds prey not only on fish of an array of species, but on squid, crustaceans, like copepods and shrimp, seabed-dwelling molluscs, such as mussels, the floating or washed-up bodies of dead sea mammals and, in some cases, each other. Some, such as the storm-petrels, are specialists at picking floating prey from the surface, while others are deep divers – the

Every summer, our sea cliffs play host to some of the most important seabird colonies in the world.

Razorbill can propel itself to depths of 120m. The most adaptable eaters are the large gulls, which can catch fish quite proficiently but (in seaside towns at least) are just as likely to lunch on the contents of a ripped-open bin bag, or a Cornish pasty expertly swiped from an unsuspecting tourist's hand.

Most seabirds are strong flyers. Several of the long-winged species use a specialised 'shearing' flight, whereby they utilise air currents generated by wave motion to glide with minimal energy expenditure. One way or another they can travel huge distances as they forage. Because their feeding grounds are so clearly separated from their nest-sites, they have little need for territoriality and only defend the immediate areas around their nests from intruders. Most therefore breed in large colonies, enjoying the many advantages of communal living. To visit one of these 'seabird cities' at the peak of the breeding season is to experience a sensory assault the equal of any wildlife 'spectacular' on Earth.

When breeding is over for the year, most of our seabirds abandon their colonies completely, and may travel thousands of kilometres away to exploit more favourable feeding grounds in the southern hemisphere. In the autumn exposed headlands around the coast of the British Isles become meccas for birdwatchers, who brave inclement weather to observe seabirds passing offshore, sometimes in huge numbers, on their southbound journeys.

Sealife vastly outweighs life on land, in both number and variety. Exploration and study of the underwater world is more difficult for us than surveying life on dry land, but it is an undertaking of crucial importance, because pollution, overfishing, climate change and physical damage to reefs and other inshore habitats are all harming sealife across the globe. Seabirds are key indicators of the state of play in marine ecosystems, because unlike most sealife they are highly visible to us, their breeding habits make their populations relatively easy to monitor and their position as high-level predators makes them early casualties in any incipient ecological collapse. When most or all chicks in an auk or tern colony die of starvation, this is a clear indication that prey species such as sandeels are undergoing a population collapse or a radical change in distribution – or both.

The Fulmar is a miniature cousin of the albatrosses, and like them travels thousands of sea-miles over its long life.

CONSERVATION OF SEABIRDS AND THEIR HABITATS

All wild birds in Britain are protected by law. It is illegal to kill them or destroy their nests. The legislation that protects them is the Wildlife and Countryside Act 1981 and its associated amendments. According to an additional layer of protection (Schedule 1 listing), it is illegal to disturb rarer breeding birds in any way while they are nesting. Several species discussed in this book are protected in this category – all species of divers, the rarer grebes, Scaup, Common Scoter, Leach's Storm-petrel, Mediterranean Gull, and Little and Roseate Terns. A few bird species may be destroyed under a general licence (which need not be applied for) if they are causing serious danger to public health or safety – this includes certain common species of gulls.

In 1969–70 a full census of Britain's seabird populations was organised by a charitable organisation called the Seabird Group, in a project known as Operation Seafarer. A follow-up census, the Seabird Colony Register, was arranged by the Seabird Group and the Nature Conservation Committee, now the Joint Nature Conservation Committee (JNCC), and took place between 1985 and 1988. The most recent full census was Seabird 2000, again spearheaded by the JNCC, which gathered data from 1998 to 2002 and employed some new detection methods (using playback at nocturnal petrel and shearwater colonies to see whether burrows were in use) to obtain the most accurate data so far. Results from these three projects give a clear picture of changes in breeding seabird populations. The JNCC has an ongoing Seabird Monitoring Programme, and plans to begin the next full census in 2015 or 2016.

In the UK a programme is underway at the time of writing to identify and protect a network of key sites at sea that are regularly used by high numbers and varied species of seabird. Measures to safeguard these Marine Special Protection Areas include limiting disturbance and pollution risk, by regulating – or in some cases banning – shipping and developments such as wind farms.

WHAT MAKES A SEABIRD?

Many birds make use of coastal habitats, either sporadically or habitually. Visit a quiet beach in winter and you may see Skylarks, Linnets and Meadow Pipits, birds of open countryside, picking through the tideline debris, alongside small birds more closely associated with the coast, such as Shore Larks and Rock Pipits. At the water's edge waders like Sanderlings and Bar-tailed Godwits search the wet sand for burrowing worms and other prey as the tide retreats. Birds of prey such as Merlins and Peregrines are attracted to these gatherings. The latter species also nests on sea cliffs, as do Ravens, Rock Doves, Kestrels and Jackdaws. It is not that unusual to see freshwater-breeding wildfowl like Mute Swans and Wigeons swimming on calm seas, and Kingfishers often head for the coast in winter when inland waters freeze over, to fish around estuaries. However, the number of species that routinely find their food either in the sea or floating on it is rather smaller.

The species defined in this book as 'true seabirds' are not casual visitors to the coastline, but are obliged by their biology to forage for most if not all of their food out at sea, at least during the non-breeding season, if not all year round. For a few of these species (the Cormorant and certain gulls), part of the population has become secondarily adapted to a life based mainly or entirely inland, usually as a result of human-made environmental changes. However, they are still regarded as seabirds and a significant proportion of their population retains a sea-based lifestyle.

The northern hemisphere's answer to the penguins, auks like the Black Guillemot are consummate underwater swimmers but have not (quite) given up the other use of their wings.

ABOUT THIS BOOK

This book describes all seabirds that occur, or have occurred, around the British and Irish coastlines. It is structured along taxonomic lines, so related species are grouped together. Each chapter focuses on a particular group of breeding species, and begins with detailed accounts of species within that group. In addition to our breeding species, a number of other seabirds have been recorded off British and/or Irish coasts. Some are regular visitors, passing our shores on their annual migrations. Others are wanderers from much further away, and have been recorded only once or a few times. These birds are covered with briefer accounts at the end of the relevant chapters. The species described in this book are representatives of the following taxonomic groups.

Opposite. Healthy undersea ecosystems are key to our holding on to iconic fish-eating seabirds like the Puffin.

Order	Family
Anseriformes	Anatidae (ducks, geese, swans)
Gaviiformes	Gaviidae (divers)
Podicipediformes	Podicipedidae (grebes)
Procellariiformes	Procellariidae (shearwaters and Fulmar)
	Diomedeidae (albatrosses)
	Hydrobatidae (storm-petrels)
Suliformes	Fregatidae (frigatebirds)
	Sulidae (gannets and boobies)
	Phalacrocoracidae (cormorants and shags)
Phaethontiformes	Phaethontidae (tropicbirds)
Charadriiformes	Phalaropodidae (phalaropes)
	Scolopacidae (waders)
	Laridae (gulls)
	Stercorariidae (skuas)
	Sternidae (terns)
	Alcidae (auks)

FAME

Some of the data discussed in this book have come from the Future of the Atlantic Marine Environment (FAME) project, which at the time of writing is still ongoing. This large coordinated project is led in the UK by the RSPB and aims to build a much more complete picture of our understanding of seabird ecology in the Atlantic (including the North Sea), using a range of methods to study seabird behaviour, breeding success, foraging range and other aspects of their ecology. To find out more visit the project's website at www.fameproject.eu/en/

Seaducks

The family Anatidae includes the swans, geese and ducks - a group familiar and clearly recognisable to most people. Most species in the group have distinctive broad, flattened bills, sturdy legs, feet with webbing between the front three toes, stocky bodies and short, pointed wings that they beat rapidly in their often very fast flight. The majority breed and spend their winters on or alongside fresh water, although they may visit salty coastal lagoons and estuaries, and some species even roost on the sea.

The species in the subfamily Merginimae are diving ducks, the majority of which can truly be classified as seabirds. They spend the entire non-breeding season offshore and feed on prey caught in the water or pulled from the seabed. These birds are generally known as seaducks, with the most highly adapted species being the eiders and the scoters. They are sturdy, powerfully built ducks that can dive to impressive depths (more than 40m in the case of the Common Eider). Unlike most seabirds they propel themselves with their feet rather than their wings when swimming under water.

Some seaduck species, such as scoters, breed well inland by fresh water, others by sheltered coastal bays. As with all wildfowl, young seaducks are very active and feed themselves from the first day of their lives. They therefore benefit from having access to reliably calm and food-rich waters in which to forage, as well as from more cover in which to shelter from predators. Outside the breeding season they are generally seen in large groups, and tend to stay close to areas with large mussel beds or other concentrated food sources for the duration of the winter. Most seaducks are very rare on inland waters, although they may take shelter on large reservoirs, for example, following storms at sea.

Like their freshwater relatives, seaducks display sexual dimorphism, the males are more colourful or boldly patterned than the females. Courtship displays, which begin in midwinter, are often communal, with several drakes posturing vigorously and calling noisily to the females. Pairs form several weeks before the birds move on to their breeding grounds in spring, but the pair bond persists only until the female has laid her eggs. She incubates and rears the young alone (as the ducklings feed themselves they do not need the same level of parental care as other seabird chicks). Post-breeding, the annual moult renders the adult ducks temporarily flightless as the primary and secondary feathers are replaced. This is not a serious handicap as they are back at sea when the moult begins and therefore have less need to fly. Males may assume a drabber 'eclipse' body plumage during the moult.

With its superlative waterproofing and insulation, the Eider has little to fear from even the roughest seas.

Eider
Somateria mollissima

A large, solid and very distinctive bird, the Eider is probably our best-known seaduck. It is also arguably our most marine duck, as it breeds on the coast as well as wintering at sea. It is famous as the source of eiderdown, the supremely soft and warm underlayer of plumage that the female plucks from her breast to use as nest lining, and this is still sustainably harvested at some breeding sites once the ducklings have hatched and left the nests. The species' scientific name references the pleasing properties of eiderdown – *somateria* means 'woolly body' and *mollissima* is Latin for 'very soft'.

INTRODUCTION

The Eider or Common Eider is a big, stocky and thick-necked duck with a distinctly 'Roman-nosed' profile, and cheek feathering that extends some way along the bill sides. These features give the bird a unique facial look, which is apparent (given a close view) in all of its varied plumages. Females are uniformly mottled brown, while breeding drakes are boldly patterned in black and white with minty-green patches on the neck and a beautiful pinkish wash on the otherwise white breast. Immature (under three years) and eclipse-plumage drakes can show confusing intermediate patterns, often with much black or very dark brown feathering.

This duck is a very strong swimmer and diver, its chunky proportions and subcutaneous fat (along with its celebrated down) help it to stay warm when immersed in cold Arctic waters. Due to its small wings and heavy body it has very high wing-loading, so it needs to expend a great deal of energy to stay airborne. Its hard-flapping flight, powered by proportionately huge pectoral muscles, is extremely fast, approaching 80kph (making the Eider one of the world's fastest-flying birds).

In winter Eiders are highly gregarious, feeding, resting and, as spring approaches, courting in flocks.

DISTRIBUTION, POPULATION AND HABITAT

This seaduck breeds around coastlines across much of northern Europe, including Svalbard, and Asia, North America and Greenland. Its world population is estimated to be 3.1–3.8 million individuals, with about 30,000 pairs breeding around the British Isles. The wintering population is about 80,000 individuals, with visitors from north-east Europe joining the local breeding birds. Eiders breed in Scotland, Northern Ireland and parts of northern England, and may be seen offshore anywhere around the coast, but are more common further north. Small non-breeding flocks may be seen offshore in summer, but the species is much more commonly seen in the open sea during winter.

Eiders nest on low, rocky coastlines in sheltered areas, especially on islands where their young can be safe from mammalian predators. It is not unusual for them to nest in the proximity of Arctic Tern colonies, where they benefit from the terns' hyper-vigilance against predators. They spend the winter months offshore, often in sheltered bays and estuary mouths, close to good food supplies. It is extremely rare to encounter wild Eiders inland. However, the species is rather popular in captive wildfowl collections and such birds occasionally escape and may turn up anywhere.

The drake Eider is a handsome bird, with a very distinctive expression thanks to its high-set eyes and 'long face'.

EXOTIC EIDERS

There are four to six recognised subspecies of Eider worldwide. Our breeding birds are of the nominate subspecies *Somateria mollissima mollissima*. The subspecies *S. m. borealis*, native to the High Arctic, has been recorded in British waters. Drakes of this form have distinctive bright orange-yellow bills, and frilly extended scapular 'sails', which are small but often quite obvious given a clear view. Eiders that breed on the Shetlands are smaller than other UK Eiders and may belong to the subspecies *faeroensis* rather than *mollissima*.

A family of Shetland Eiders.

Communal courtship is a chaotic and noisy business, with displaying, chases, scuffles and constant outraged-sounding 'ah-OOOH!' calls.

BEHAVIOUR AND DIET

This is a very gregarious duck, usually encountered in flocks and quite often in gatherings of more than 1,000, especially in winter. In the summer months birds that are too young to breed stay at sea in smaller flocks. Flocks tend to stay closer inshore in calmer waters to rest in between feeding sessions (for example when high tide makes the seabed less easily accessible). When moving between feeding and resting spots they fly very low over the water in a series of long lines.

When actively feeding, groups of Eiders dive in waves, with certain birds acting as leaders to the rest. Studies of groups of females within flocks suggest that the birds that initiate feeding and serve as leaders are usually those with the poorest body condition, and therefore presumably have the greatest need to feed. Followers benefit from the leaders' eagerness as they can make more efficient dives, heading directly to where the leaders have found food.

Eiders feed primarily on small marine animals found on the seabed, especially molluscs, sea urchins and crabs. The single most popular prey is the Blue Mussel *Mytilus edulis*. To access and retrieve this prey, the ducks need to make long and deep dives. They are foot-propelled divers, like other diving ducks, although underwater footage shows that they may beat the half-closed wings at the start of the dive to assist in gaining depth more quickly. They can reach depths of 40m, although typical foraging depths are around 10m. Once at the seabed, they use dabbling motions of the bill to search for prey in the substrate. The prey is usually brought to the surface to be eaten; even quite large and uncomfortable-looking items like spiky sea urchins can be swallowed in one go, although the birds usually bite and shake off the long legs and claws of spider crabs before swallowing the bodies. Eiders may also feed in shallow water where they can access the seabed by upending or even just immersing their heads, and in the shallows they may use their feet to dig into the seabed substrate. In winter they tend to seek out larger prey in order to reduce time spent under water, since diving in colder water is more energetically expensive.

Digesting prey like mussels and sea urchins, which have very hard body parts, requires some serious internal processing. The gizzard of a bird, part of the stomach, is a very tough and muscular organ, with vigorous peristaltic action that in seaducks crushes up crustacean carapaces and mollusc shells so that they can safely pass through the digestive tract. Very young ducklings stick to soft-bodied prey such as amphipods.

BREEDING

Eiders form pair bonds during the late winter, when in large mixed flocks. Adult drakes (those that are at least three years old) display communally, with several often surrounding a single female, each rearing up in the water and tossing its head back repeatedly. The display is accompanied by a distinctively fruity and comical 'aaah-ooo' call. Once the pair bond is formed the male remains close to his mate for several weeks, and there is evidence that the protection of an attentive mate allows females to feed for longer, as they are not being distracted by the advances of other drakes. Since the female is able to access extra food, she is in better condition at the start of the breeding season than she would be without the protection of the male, and consequently is more likely to be successful in breeding. She may lose 40 per cent of her body weight over the duration of incubation, so preparation is very important. By sticking close to his mate, the male also improves the chances of ensuring that he is the father of her brood – but his help is not required for the actual care of the eggs and ducklings, so once the eggs are laid he normally moves on and joins other males in flocks to resume the usual communal lifestyle.

The down that lines an Eider nest provides cushioning and superb insulation for the eggs.

Female Eiders can breed from the age of two years, and may live for more than 30 years. It is not unusual for adult females to make no breeding attempts in some years. This is quite a common strategy in long-lived birds. Missing a year is most likely when a female is not in peak condition at the start of the breeding season. Opting not to nest (and thus avoiding all the physical stresses and risks that nesting entails) in that year gives the bird extra recovery time for the following year, and may allow her to increase her overall lifetime reproductive potential. Females with low body condition are also more likely to abandon their newly hatched chicks than are those in peak condition – however, the young do at least stand a chance of survival, as they may be able to join another family.

Females nest in low, sheltered hollows close to calm seas, often among boulders or vegetation. They frequently form loose colonies, with just a few metres separating the nests. Bolder and more experienced females tend to nest further from the shoreline than shyer birds – this willingness to venture inland allows them to find more sheltered sites. The selected hollow is lined with eiderdown, exceptionally soft and fluffy feathers that the female pulls from her breast and pats or treads into place. She lays her eggs (usually four or five, but can be up to eight) at daily intervals, but does not begin to incubate them until the clutch is close to complete. She continues to

The Farne Islands in Northumberland are a good place to observe Eider nesting and family life.

Predators have already made significant inroads on this Eider crèche; the two females would originally have had eight or more ducklings between them.

add down to the nest throughout the 25–28 days of incubation. The heat-holding properties of the down keep the eggs sufficiently warm at times when the female has to leave the nest to feed. The eggshells are yellowish or greenish-brown to camouflage them from predators. Nevertheless, they are still highly vulnerable when unattended, and the female's breaks from incubation are few and brief. Her own camouflage is superb, and she is likely to sit tight on the eggs until the last possible moment if a predator wanders near the nest.

Like the young of all ducks, the young are precocial – able to move freely and feed themselves very soon after hatching. They have a coating of down in a camouflaged brown pattern, which quickly dries off once they are out of the egg, and in less than an hour they are strong enough to run and swim after their mother. She leads them to the sea, a safer place for them than the land, although there are still many dangers. The female does her best to guard them and drive away would-be predators – the large gull species such as Herring and Great Black-backed Gulls present the most significant danger. Often several Eider families team up once on the water. By living in these crèches, all benefit from a larger 'attack force' – any predator may be set upon by multiple female Eiders, which will fearlessly launch themselves air-wards and attack a cruising gull or skua. The ducklings can also protect themselves by diving. However, predation rates may still be very high – in some areas more than 90 per cent of ducklings are taken by gulls. Some individual gulls may be specialist predators of Eider ducklings. Some Eider populations in areas with few local gulls have much higher survival rates.

The young birds are most vulnerable to predation in their first two weeks of life. Ducklings are more likely to be taken in rough rather than calm weather, probably because it is more difficult for them to stay close to their mother in rough conditions. Small ducklings are also vulnerable to the direct consequences of bad weather – deaths from exposure are common. The ducklings become significantly less vulnerable to both predation and the effects of bad weather once past two weeks of age, and some of the females in the crèche group start to move away at this point. However, the ducklings still benefit from the protection of at least a few adults for another five or six weeks. They are able to fly at about 65 days old, and thereafter both the youngsters of the year and the breeding females join flocks at sea. Young Eiders have, on average, a 33 per cent chance of surviving their first year, but survival rates for adults are much higher.

CROSS PURPOSES

There are documented records of hybridisation between the Eider and the closely related King Eider. Wildfowl as a group are well known for their readiness to hybridise, so it is no surprise that other probable hybrids between the Eider and more distantly related species have also been observed, such as Eider × Red-breasted Merganser and even Eider × Shelduck.

MOVEMENTS AND MIGRATION

The Eiders that nest around the British coastline tend to spend their winters fairly close to their breeding grounds – it is unusual for them to move more than 200km from the coastline where they breed, although there is substantial variation between colonies. For example, adults that nest at the Sands of Forvie in Aberdeenshire mostly move south to rich feeding grounds in the Firth of Forth or Firth of Tay, well over 100km away. On the other hand, the Farne Islands birds do not make a similar-length journey northwards, instead remaining very close to the islands throughout the winter. Numbers of Eiders in UK waters are augmented by migrants from more northerly coasts, some of which have to move south for winter as the seas around their nesting grounds freeze up in the coldest months. Certain 'hotspots' around the coast are used regularly by moulting flocks.

Some British-born Eiders do travel substantial distances. There have been recoveries of birds ringed in Britain from several northern European countries. One male ringed as a chick in eastern Scotland was found 1,212km away in Uppsala, Sweden. The longest southbound journey was that of a 24-year-old bird ringed in Fife and recovered 750km away in northern France.

Although fast and strong in flight, the heavy-bodied Eider needs a good run-up to achieve lift-off.

THE FUTURE

The Eider has the distinction of falling under one of the earliest ever bird protection laws, with the population nesting on the Farne Islands in Northumberland being protected under a law established by St Cuthbert in the 7th century. The species is still known locally as Cuddy Duck in honour of the saint. Today, Eiders still have full legal protection. However, they may be shot under licence if they are shown to be causing damage to mussel farms (provided non-lethal measures have been tried first). In some other European countries Eiders are legal quarry.

The conservation status of the Eider in the UK is Amber, reflecting the importance of the wintering population here, which according to the most recent surveys is in a long-term gradual decline. The breeding population, however, is now stable following around 200 years of steady increase. The situation is the same in the case of the Irish population. The exception is the Shetland population, which has declined steeply in recent years; numbers fell from 16,500 birds to 6,000 between 1977 and 1997. The IUCN's assessment of the species is Least Concern, reflecting its very large and broadly stable global population, although (as is the case with the Shetland birds) some local populations are in steep decline, while others are increasing.

Like other seabirds that spend most of their time on the surface of the water, Eiders are very vulnerable to oil spillages and other pollution incidents. The Eider was one of the species most severely affected by the *Braer* disaster in Shetland in 1993, while a spill in the Firth of Tay in 1968 killed an estimated 50 per cent of that area's wintering population. Conflict with mussel farmers is another potential hazard, although there are many non-lethal options farmers can take to reduce the impact of Eiders and other seaducks. Eiders breeding on the mainland may be at risk from non-native predators such as the American Mink, and some populations may suffer as a result of the growing population of Great Skuas (although other important predators of Eider ducklings, such as Great Black-backed Gulls, are declining). Changes in food availability may also have significant impacts on population levels. In areas where natural mussel populations have been over-harvested, such as the Wadden Sea, there have been dramatic declines in Eider numbers.

Overall, the Eider seems to be in a reasonably secure position as a UK breeding bird, so declines in winter numbers may be due to reduced numbers of visitors from further afield. Nearly 10,000 Eiders regularly winter across several Special Protection Areas, and many breeding sites are also within protected areas.

Common Scoter *Melanitta nigra*

Look well out to sea across any sheltered bay around our coastline in winter and you stand a chance of seeing a few, or perhaps a lot, of these handsome dark seaducks. At a distance they may look as though they are packed together in a tight raft. Getting a closer look at them is quite challenging because they are often some distance offshore, especially when feeding. As breeding birds Common Scoters are extremely scarce and declining in Britain and Ireland, where they breed some distance inland on very remote moorland. However, wintering numbers exceed 100,000, as visitors from further north and east move into our seas. It is always worth scrutinising these flocks carefully, as they may include other, rarer scoters. The Common Scoter is on the Red List of species of conservation concern in the UK because of its declining breeding population.

INTRODUCTION

This seaduck is considerably smaller than our other common seaduck, the Eider, and is a little smaller than the most familiar freshwater duck, the Mallard. The male has entirely jet-black plumage with a slight gloss. The bill is black with a little yellow along the centre of the upper mandible, and a swollen 'knob' at its base. Females and juveniles are dark sooty-brown with paler (but still very dusky) cheeks, rather like a dark version of the female Red-crested Pochard, and lack the bill knob.

The bird's shape is typical for a seaduck, with a sleek outline, heavy, deep-chested body that sits quite low in the water when swimming, and rather long, pointed tail that may be held cocked upwards. Adapted to diving in cold water, the species has a high body weight and small wings for its size; it therefore struggles to get and stay airborne, and flies with strong and rapid wingbeats. However, they are quite capable of flying considerable distances, and breeding sites may be many kilometres inland. Common Scoters are also more likely to turn up inland in winter than Eiders.

The drake Common Scoter's plumage has no white areas, making it easy to distinguish from Britain's other regular scoter species.

DISTRIBUTION, POPULATION AND HABITAT

The Common Scoters that breed in the British Isles are found in northern Scotland and the western side of Ireland (where breeding was first recorded in 1905). The species has always been a scarce breeding bird here but its numbers have halved since the mid-20th century, and there are now thought to be 50–60 nesting females, distributed very patchily across a wide area. In winter Common Scoters, mostly of continental origins, can be seen around the entire coastline, although the largest flocks occur in more northerly and sheltered areas. Important concentrations gather in the Moray Firth, Firth of Forth, Liverpool Bay, Carmarthen Bay and off the North Norfolk coast. Non-breeding birds may be seen offshore in the spring and summer, although in much smaller numbers.

The British breeding population is very much an outpost – Common Scoters breed in a broad band across Norway, northern Sweden, the Baltic countries and much of northern Russia. A full assessment of the world population is difficult as the breeding habitat is inaccessible, but the IUCN estimates there to be 1.6 million individuals worldwide. These birds move south and west in winter to spend the colder months offshore along the coast of northern Europe and down along the west coasts of France, Spain and north-west Africa, with a few occasionally wandering to the Mediterranean.

Breeding habitats are located around quiet inland lochs and lochans in remote upland moors with low, heathy vegetation. For example, the Flow Country in Caithness is a typical nesting habitat, although the birds seem to select certain lochans and avoid other similar ones for no obvious reason. Many subtle factors, such as water depth, elevation and the presence of trout, may influence the abundance of small invertebrate prey in the water, which is food for both adults and ducklings and therefore key to whether breeding will be successful. Migrants travelling to or from breeding grounds may briefly visit inland waters. Winter habitat is sheltered inshore waters, and occasionally stray individuals or small parties visit inland lakes and reservoirs. Birds that turn up inland in winter are likely to have been storm driven, but if the water offers good feeding opportunities they may stay for some weeks.

Until recently the very similar Black Scoter of North America was considered to be a subspecies of the Common Scoter, but now most authorities split this form as a full species, leaving the Common Scoter as a monotypic species. Black Scoters have been recorded on occasion in UK waters, and are discussed at the end of this chapter (see page 29).

BEHAVIOUR AND DIET

Although scoters and other seaducks typically dive to find prey, some species have been observed to take exposed mussels at the surface.

Common Scoters are highly gregarious, and their flocks may include other scoter species, especially Velvet Scoters. They move between feeding spots and more sheltered areas to rest, although they are rather active at all times, often taking off en masse to fly just a short distance, and when sitting on the water they frequently rear upright to stretch and flap their wings. When flying a longer distance a flock arranges itself into long, straggling lines that move low over the waves.

Like other seaducks Common Scoters are capable of deep dives, and when at sea they readily dive up to 20m down to reach prey on the seabed. The state of the tide influences how easily they can access mussel beds, and thus dictates their activity pattern through the day (and night). When feeding many or all birds in a flock dive in quick succession, beginning the dive with a small jump. Once under water they descend quickly with powerful strokes of their broad feet to reach the seabed, where they probe and push into the substrate with their broad bills to locate prey and (if necessary) pull it free. The bill edges have marked serrations to improve grip. Mussels, especially the Blue Mussel *Mytilus edulis* and other bivalve molluscs, are favoured food, but they also take gastropod molluscs and sometimes small fish. The prey is brought to the surface to be eaten.

A female Common Scoter collects preen oil from her uropygial gland. Good waterproofing and insulation is essential for this deep diver.

While winter rafts of Common Scoters are easily found, the females on their breeding grounds are much more discreet in their habits. Feeding behaviour in the lochs on the breeding grounds is rather different as well, as the birds forage in much shallower water – usually under a metre deep and close inshore. However, adults and chicks alike still mainly forage by diving rather than upending or head dipping. Aquatic insect larvae make up an important proportion of the diet of breeding birds, and fish are taken more often inland than at sea.

BREEDING

Within the large winter flocks, males and females increase their interaction as spring approaches. Males gather around a female to display, adopting upright postures, cocking and fanning their tails, simultaneously bowing and flapping, or rushing towards the female in a low stance with their neck extended. The female may respond in kind with head bowing and other posturing of her own. As all this is going on, the birds make various whistling and croaking calls.

With no colourful plumage to show off, the male Common Scoter's courtship display makes much use of exaggerated body postures.

In the weeks before egg laying, pairs stay close together within the flock. The male's guarding behaviour possibly allows easier foraging for the females, as other drakes are kept away. Observations of pairs on the breeding grounds suggest that females do not spend extra time feeding compared with males, so it is likely that the female's important 'feeding-up' phase to maximise condition for breeding takes place primarily at sea, just before the move inland.

Paired-up couples migrate together to their breeding grounds in April or early May. They may also be joined by unpaired birds of both sexes that are looking for a last opportunity to find a mate and breed. Males that do have mates therefore need to stay with their partners and guard them for a little longer, up until the females begin to lay their eggs. Then the males depart, heading back to the sea to join flocks of other males that have completed their breeding duty. The young and other non-breeding birds of both sexes also head to sea at this time.

A female ready to lay selects a well-concealed hollow for her nest, close to the water's edge but with plenty of sheltering vegetation around, and ideally on an island in a loch or lochan. During incubation both she and the eggs are highly vulnerable to predation – in areas where the non-native American Mink occurs, this mammal can have a very significant impact on nesting success. Foxes and Pine Martens are also key predators of both eggs and chicks, as are Common Gulls. Where the scoters share breeding lakes with Black-throated Divers, the divers are also likely to kill ducklings, although they are less dangerous to the eggs.

The clutch consists of six to eight eggs, which are laid on consecutive days. Their shells are whitish and thus highly visible during the female's breaks from incubation if the nest hollow is not well concealed. She leaves the nest infrequently and her feeding trips are brief. While she is away, the nest's lining of downy feathers helps to conserve heat. Incubation takes about 30 days, and begins when the clutch is nearly complete, so that the ducklings all hatch in quick succession. They are ready to follow their mother to water very soon after hatching, and join her in the shallows to dip and dive after swimming insect larvae and other prey. Common Scoter females are not particularly inclined to nest in close proximity to each other, but if they do then two or more broods may join together.

The ducklings take about seven or eight weeks to mature enough for their first flights. They are highly vulnerable to predators when small, but those that survive their first two weeks of life greatly improve their chances of reaching maturity. Soon after they become capable of flight, they and their mothers leave the breeding grounds and head for the sea, joining existing flocks. The young birds, which are almost indistinguishable from adult females by this point, have an extended period of sea-faring life ahead of them, as they will not normally breed until their third year. This species can live for more than 15 years (possibly for much longer – there is little data available), so it is possible that females adopt a similar strategy to Eiders and only breed in years when they are in peak condition, but this bird has been much less studied than the Eider.

MOVEMENTS AND MIGRATION

After breeding, females with the young migrate to the sea, where they join adult males and non-breeding birds that are already there. The annual moult then takes place – for some weeks the birds are flightless as they drop and regrow their primary and secondary feathers, and they need the safety of the open sea during this vulnerable time. The moulting flocks tend to disperse somewhat once the birds are fully flighted again, although some stay in the same areas throughout the winter.

Some birds make rather short journeys and overwinter at the nearest coast to where they nest, but others travel many thousands of kilometres. There have been few ringing recoveries for this species, but they include recoveries in the UK of birds ringed in Iceland, Estonia and Russia. One bird recovered in Russia in 2007 had been ringed during its rehabilitation after being caught up in the 1996 *Sea Empress* oil spill in south Wales, demonstrating that rehabilitation of oiled scoters can be entirely successful.

Scoters of all species can cover long distances over the sea, although most birds remain inshore in winter.

THE *SEA EMPRESS*

Like other seabirds that spend much time resting on the surface, Common Scoters can be highly vulnerable to oil pollution, should a spill occur at one of the particular areas where large numbers congregate. Exactly this happened in February 1996, when the *Sea Empress* shed 72,000 tonnes of crude oil into Carmarthen Bay, the winter home of some 16,000 Common Scoters – 30 per cent of the UK's whole wintering population. A total of 4,571 scoters were picked up dead or dying, and no doubt more were undiscovered – 83 per cent of all oiled birds found after the incident were Common Scoters. Several hundred birds were taken to wildlife rescue centres for oil removal and rehabilitation, although subsequent recoveries (the birds were ringed under licence after treatment) suggest that at least 10 per cent of these died soon after release.

A casualty of the Sea Empress *is cleaned up by helpers.*

THE FUTURE

Common Scoters are declining in the British Isles as both breeding and wintering birds. The breeding population here is very small, and makes up a tiny fraction of the world's total. The arrival of American Mink and their predation upon scoter chicks is a key factor behind the loss in the 1990s of an important (and only recently established) population at Lower Lough Erne in County Fermanagh. This was Ireland's first site to hold the species and at its peak was home to up to 180 nesting females. Over-eutrophication of breeding lochs is a further contributor to declines – this can cause considerable unbalancing of the aquatic ecosystem, allowing trout populations to increase and thus compete with scoters for available insect prey. Climate change is another factor, with the optimum climate range for the species being pushed north and shrinking year on year. To protect the remaining breeding birds, it is important that we develop a clearer understanding of exactly what factors limit breeding success.

There are other threats besides oil pollution that affect wintering Common Scoters at sea. Shipping activity can drive scoters away from their preferred feeding grounds, and it is possible that offshore wind farms could also cause disturbance. Common Scoters may come into conflict with mussel farmers, although probably not to the same extent as Eiders. Although the species is fully protected in the UK (under Schedule 1, which gives it additional protection from disturbance at the nest), it is legal quarry and its eggs are harvested for food in some other countries, which may affect numbers reaching British waters in winter.

Without coordinated efforts, Common Scoters off our shores could become a rare sight.

Providing effective conservation for this bird involves managing multiple factors. The RSPB has undertaken an extensive study of the Common Scoter's breeding ecology, to help identify the best sites for breeding productivity and the key factors that need to be managed to improve breeding success. Protecting the wintering birds at sea involves minimising disturbance at their favoured areas, and improving technology to reduce the chance of future oil spills.

Red-breasted Merganser
Mergus serrator

The Red-breasted Merganser is the most marine of the three 'sawbill' ducks that occur in Britain. It is a sleek, striking and sometimes rather comical-looking bird, and is a specialist hunter of fish, which it secures using the serrated inner edges of its bill.

INTRODUCTION

The sawbill ducks are all very distinctive, with their slender, tooth-edged bills and crested heads, and the Red-breasted Merganser has the slimmest bill and most outrageous crest of them all. The drake is rather colourful with his green-glossed head, white collar and chestnut breast, and both sexes have very long, fine crests extending from the back of the crown and the nape, producing a distinct 'punk hairstyle'. In flight this species shows prominent areas of white in the wings, and looks slim and attenuated, rather than thickset and sturdy like a scoter or eider.

The fabulous spiky head plumage of the Red-breasted Merganser gives it a striking silhouette.

The dusky neck and breast of the female Red-breasted Merganser helps distinguish it from the more neatly marked female Goosander.

DISTRIBUTION, POPULATION AND HABITAT

This bird is a fairly rare breeding species, with a strong northern/western bias to its British distribution. It increased in its Scottish and Northern Irish stronghold throughout the late 19th and early 20th centuries, first breeding in northern England and Wales in the 1950s. There are in the region of 2,500 breeding pairs in the UK and a similar number in the Republic of Ireland. Most of these birds spend the winter locally, but numbers in British waters are increased by influxes of birds from northern Europe. The species breeds across most of northern Eurasia and North America, and has a population of over half a million. In Britain it nests inland by slow-flowing rivers and lakes, and also along sheltered coasts. During the winter it is usually at sea, using sheltered bays and other calm inshore areas, and also sometimes visits inland water bodies.

BEHAVIOUR AND DIET

Red-breasted Mergansers are fairly sociable but tend not to form the huge gatherings of scoters and eiders, and it is common to see singletons and small parties. When swimming they have a very low profile, and often dip their heads under water in the manner of grebes, looking for prey before they dive. They are fast and agile under water, as they need to be able to catch fish, and their dives can last for more than 30 seconds. However, this species is not typically a deep diver, rarely going below 5m. It brings each caught fish to the surface to eat it, and at this point may be targeted by opportunistic gulls that attempt to steal its catch. It preys on whatever fish species are available – on the breeding grounds these are usually Brown Trout, Grayling and small salmon, while at sea it takes herring and many sandeels. It also sometimes feeds on swimming crustaceans, and when breeding the adults and chicks both take aquatic insect larvae.

Like most ducks, Red-breasted Mergansers perform communal courtship displays, males neck-stretching to show off their white collars.

BREEDING

The courtship and pairing behaviour of the Red-breasted Merganser is fairly typical for ducks, with pairs forming during the winter and persisting up until the female lays her eggs. In courtship drakes perform a bowing and neck-stretching display around a female, sleeking their fluffy crests into a point and calling. This posturing is interspersed with rapid dashes across the surface.

The pair moves to the breeding grounds in March, and the female selects a suitable sheltered hollow close to the water's edge, on the ground or among boulders, or in the form of a hole in

Females search among vegetation close to water for a suitable hollow in which to nest.

a tree, in which she lays a clutch of eight to ten eggs. She will readily use an artificial nestbox if available. The eggs hatch after 31 days of incubation, and the family then takes to the water, sometimes teaming up with other broods to form crèches. As a result of this arrangement, when the ducklings have outgrown their most vulnerable stage some of the adult females can depart, joining the males at moulting sites, and leaving just one or two females in charge of several broods of ducklings. When they are two months old the young mergansers are fully independent and move away to spend the winter months at sea.

A few British Red-breasted Mergansers range north and east as far as Finland in winter, but most stay local.

MOVEMENTS AND MIGRATION

Elsewhere in its range the Red-breasted Merganser is quite a prodigious migrant, with some Asian birds travelling thousands of kilometres to winter on the Arabian Sea. However, most of the British breeding birds winter at the closest suitable offshore point to their breeding grounds, and winter immigrants to British waters are unlikely to have come much further than from northern Europe. There are a few records of birds ringed in Britain and recovered in Finland, Norway and Denmark.

THE FUTURE

The breeding population is showing signs of decline in Scotland and Ireland, although numbers are stable in Wales and England. One concern is that species' breeding habitat may be vulnerable to accidental disturbance. Its dietary habits have brought it into conflict with angling interests, and fishery owners may obtain a special licence to cull it where it is considered to have a significant impact, and non-lethal measures have failed. Although it is otherwise fully protected by law, it is also sometimes illegally killed.

Other seaducks

A birder who takes the time to carefully scrutinise a raft of Common Scoters offshore stands a chance of finding something rarer lurking among them. The most likely candidate is the **Velvet Scoter *Melanitta fusca***, and it is most likely to appear off north-east coasts. This scoter is a chunky seaduck; the male is primarily black and the female is blackish-brown. At any distance it is very like the Common Scoter, although it is a little larger than the Common. The male has a mainly yellow bill and white eye-flash, and the female has more restricted pale markings on the head. The most obvious feature by far, however, is the white panel across the secondary feathers, which is present in both males and females, and revealed in flight and when the bird shakes out its wings while resting on the sea – something all scoters do rather frequently.

About 2,500 Velvet Scoters spend their winters off the British coast, travelling here from their breeding grounds in northern Europe. They are distributed between several favoured sites. The Firth of Forth, for example, attracts several hundred in some years, and in late winter in particular there may even be more Velvets than Commons present. Further south there may be just one or two among a thousand or more Commons. Their feeding behaviour is very similar to that of Common Scoters, and they may be seen performing courtship displays in late winter. Occasionally a handful of birds oversummer in Britain, but there are no breeding records.

Two pale spots on the female Velvet Scoter's face, plus white wing-bars, distinguish it from female Common Scoters.

It is worth checking through scoter rafts, as occasionally they attract a vagrant Surf Scoter, like this male.

This scoter has the dubious distinction of being one of the few British birds classified as globally threatened – the IUCN rates it as Endangered. This reflects a rapid population decline, as evidenced by numbers recorded passing key European watchpoints on migration – counts suggest a decline of around 60 per cent since the winter of 1992–1993. The world population is distributed across northern Europe and western Siberia, and is now estimated to comprise fewer than 400,000 birds.

The species faces a plethora of threats on both its breeding and wintering grounds. Like the Common Scoter it is vulnerable to oil spills and other pollution, and some breeding populations have been hit hard by the introduced American Mink. Many breeding sites have been affected by development, lake drainage and disturbance from tourism, and other threats at sea include disturbance from wind farms, as well as conflict with marine mollusc-harvesting interests. Although the British wintering birds represent a small fraction of the world population, their numbers here have remained fairly stable, so their importance may be considerable. Careful conservation of favoured feeding areas will help improve the long-term fortunes of this endangered species.

The other scoters that may be found off British coasts are much scarcer. The **Surf Scoter *Melanitta perspicillata*** is a North American species that visits British waters in small numbers – about 20 are seen on average each winter. It is intermediate in size between the Velvet and Common Scoter. The male has a white forehead and nape, which can be noticeable even from some distance away, and a rather swollen-looking bill with a bold orange, black and white pattern – he also has white eyes. The female has pale cheeks and the same swollen bill shape as the male, but not surprisingly males are noticed more often than females. Most records are off Scottish coasts, with individuals probably making return visits to the same sites in successive winters.

Common and Velvet Scoters both have very closely related 'sister species' native to North America, and both of these have been recorded in British waters. The British Ornithologists' Union split the **Black Scoter *Melanitta americana*** as a separate species from the Common Scoter in 2005, a move that prompted many keen twitchers to travel to Llanfairfechan, North Wales, to see the male Black Scoter that had overwintered there with Common Scoters. Besides this bird there are a handful of other records, mainly from Scotland. The species is almost identical to the Common Scoter, but males show much more yellow on the bill. The **White-winged Scoter *Melanitta deglandi***, North America's equivalent of the Velvet Scoter, was officially added to the British List in 2013, following acceptance of a record of a subadult male from north-east Scotland in 2011.

Vagrant King Eiders often return year on year to join the same Common Eider flock. The spectacular males are easy to identify, but females are very similar to female Common Eiders.

Finding a male **King Eider *Somateria spectabilis*** among a flock of Common Eiders is a rare treat, especially if the bird is an adult male with its resplendent blue head and bulbous-based orange bill. It is also a little smaller than the Eider, with a black rather than white lower back. Females are much more similar to female Eiders, so are less likely to be detected – on average, two or three King Eiders in total are found each winter. They arrive here from the Arctic of both North America and Eurasia – most British records are from the far north of Scotland.

Another spectacular seaduck that has been recorded in Britain, although far from frequently, is the stunning **Steller's Eider *Polysticta stelleri***. Males of this species are creamy-coloured with bold black markings, while females are dark brown with a scaly pattern. There are only about ten British records of this small, eider-like duck, mainly from the far north of Scotland, and as the species has a small and declining population (it is classed as Vulnerable), it is unlikely to turn up in any numbers in the future. It breeds in eastern Siberia and Alaska, and migrates south for the winter, with large numbers occurring on the Baltic Sea. Another dazzlingly beautiful and extremely rare wanderer from the Arctic is the **Harlequin Duck *Histrionicus histrionicus***, males of which are dark velvety-blue with white-and-burgundy markings. This small seaduck breeds across the Arctic and, unlike Steller's Eider, is currently increasing in population.

Internationally threatened and a very rare visitor to our shores, the Steller's Eider is a much sought-after species for British birdwatchers.

The mostly white winter attire of male Long-tailed Ducks is arguably the most attractive of the species' many plumage variations.

One of the prettiest of all seaducks is the **Long-tailed Duck *Clangula hyemalis***. This is a small seaduck with highly variable plumage – both sexes show distinctly different patterns in summer and winter, and have a complex moult sequence whereby some feathers are replaced three or four times over the course of a year, producing an array of transitional variation. The winter plumage is whiter than the breeding plumage, particularly in males, which are especially beautiful in their mostly white winter attire. However, the species is unmistakable in all of its guises. It is a compact, small-billed bird with white, dark grey and black tones; it has a white underside and eye-circle in all plumages, and adult males sport greatly elongated central tail feathers (although these are not always visible when the bird is swimming as they may trail below the water level).

This seaduck occurs around British coasts in winter in quite substantial numbers – some 11,000 individuals. It is most likely to be found off northern and eastern Scotland, but a few straggle to southern England, where they often turn up on coastal reservoirs and gravel workings. The breeding range extends around Arctic North America (where the species is known as the Oldsquaw) and Eurasia, with Iceland and Norway holding our nearest breeding birds. A few non-breeding birds occur throughout the summer off Scottish coasts.

The species can be found in the same sheltered, inshore seas that are used by other seaduck species, although it copes comfortably in rough seas and often forages further out than most other seaducks. When feeding it makes frequent deep dives to look for molluscs on the seabed, and can dive up to 60m in search of prey. Courtship behaviour may be observed in British waters – this is fairly typical, with males gathering to circle a female and performing ritualised head-bobbing displays. They also rush at her across the water, with their long tails cocked up.

Although their plumage varies through the year, female Long-tailed Ducks invariably have pale eye-surrounds and dusky cheeks.

Like the Velvet Scoter, the Long-tailed Duck is in trouble, with Baltic Sea watchpoint counts revealing a very rapid decline of about 65 per cent over just 15 years. It does still have a very large world population, estimated at around 6.5 million, but the rate of decline is worrying enough for the IUCN to assign it the status of Vulnerable. Large areas of its tundra breeding habitat are being destroyed via drainage and peat extraction. At sea it is at risk from pollution and entanglement in fishing nets – it is also hunted in some areas. Breeding productivity appears to be very low, and the likely cause of this is poor breeding condition in the adult females rather than increased predation of ducklings, but more research is urgently needed.

Diving ducks of the genus *Aythya* are predominantly inland species, and include common and familiar parkland species in the form of the Tufted Duck and Pochard. One *Aythya* species does have a fair claim to seaduck status, however, and this is the **Scaup *Aythya marila***. This is a medium-sized, very sleek and dapper duck. The male has a silver-grey body with a black rear end, chest and head, the latter being glossed green. The female is brown with scaly grey flanks and a prominent white patch or 'blaze' around her bill-base; both sexes have yellow eyes. This is a strong diver that feeds mainly on bivalve molluscs that it finds on the seabed.

The Scaup is an extremely rare breeding bird in Britain, with just a handful of females nesting – or attempting to nest – in northern Scotland, and there are no breeding attempts at all in some years. Worldwide, however, its breeding range is extensive, crossing northern Europe, Russia and much of North America (where it is known as the Greater Scaup, to distinguish it from the Lesser Scaup, which also breeds in North America). Some of the Icelandic and other northern European birds move to British waters for the winter, mainly down the east coast. In total about 11,000 birds are present here in winter, mostly in sheltered seas although not uncommonly on large inland waters, especially close to the coast. The global population is estimated to be 1.2–1.4 million individuals, and the overall population trend is of decline, although this is not currently severe enough for the species to be classified as threatened.

In the UK the Scaup has a Red status of conservation concern, reflecting steep declines in the winter population. Up until the early 1980s some 30,000 individuals regularly wintered in the Firth of Forth, but numbers crashed following the installation of new sewage-treatment works – however, the exact nature of the ecological change that must have occurred is unclear. Conversely, numbers in the Solway Firth on the other side of Scotland showed a strong increase when commercial cockling operations began there. Perhaps the birds were exploiting an abundance of extracted then discarded undersized cockles. The area is no longer used for this purpose and Scaup numbers have fallen again. Human activities at sea clearly have a very significant influence on the fortunes of Scaup in Britain.

Scaup are at home both offshore and on inland waters. Often seen alone, they are also highly social around rich food resources.

The uniquely patterned Shelduck has a rather goose-like outline. They are often seen in pairs, the male's large bill-knob distinguishing it from the female.

It is not unusual to see geese and dabbling ducks of various species swimming on the sea close inshore, but two species in particular are closely associated with marine habitats, although it would perhaps be a bit of a stretch to call them 'seabirds'. The **Shelduck *Tadorna tadorna*** is a large and striking duck, mostly white with a broad chestnut breast-band, green-glossed dark head, and black 'shoulder braces' and flight feathers. It is common on sheltered, mud-rich estuaries in summer and especially winter, where it dabbles, wades and upends in the shallows to reach seaweed under water. The **Brent Goose *Branta bernicla*** is a small dark goose that winters on British coasts, often feeding in large groups on the shore or in the shallows. Very large flocks arrive through the autumn, flying in untidy 'clumps' or straggling lines rather than the neat Vs of larger geese. Two subspecies visit Britain in large numbers – dark-bellied birds *B. b. bernicla* come from Siberia to winter in England, while those of the pale-bellied form *B. b. hrota* occur in Ireland in winter, and breed in Greenland and Spitsbergen.

Harbours, bays and river estuaries offer calm waters for Brent Geese to rest and feed.

Divers and grebes

The divers (family Gaviidae) and grebes (family Podicepidae) were once thought to be very closely related, a reasonable assumption given how similar they are. Both are streamlined, almost tailless, rather long-necked and dagger-billed birds with feet set far back on their bodies, making them ungainly on land. Both dive strongly in pursuit of live prey (mainly fish), and propel themselves with their feet while under water. Other traits they share are colourful breeding plumage, which is the same in the male and female, an elaborate courtship display and a strong pair bond. However, genetic study indicates that they are in fact not very closely related at all – the divers are part of a group that includes penguins and pelicans, while the grebes' closest living relatives are the flamingos. Therefore the many anatomical and behavioural similarities between divers and grebes are the result of convergent evolution – as they evolved towards similar lifestyles, so they developed similar attributes.

The two diver species that breed in Britain nest inland, as do all the grebes. Divers winter primarily at sea, but grebes show more diversity in their preferences, even within the same species. Worldwide, the divers are a small family, with just five species, all in the same genus, *Gavia*, all native to the northern hemisphere and all on the British List. The grebe family is much more diverse, with 19 or so species, representing six genera. A further three grebe species have become extinct since 1970. One of these was flightless, as are two extant (but highly threatened) species.

Diver and grebe chicks are semi-precocial, in that they are covered with down on hatching and can walk and swim on their first day of life, but need to be fed by their parents for several weeks. While ducks pair just until their eggs are laid, grebes and divers pair for the whole breeding season, and sometimes for life, as both parents are needed to take care of the chicks. As a result mate selection is taken very seriously, and their courtship displays are prolonged and elaborate, both partners playing an equal role in the performance.

At sea divers and grebes are not as gregarious or as closely tied to particular sites as most seaducks, reflecting the fact that their prey is more mobile. However, large gatherings do sometimes occur.

The elaborate courtship dance of Great Crested Grebes is one of our most celebrated natural wonders.

Red-throated Diver
Gavia stellata

It can be tricky to pick out a Red-throated Diver in its drab winter plumage against a rough grey January sea. However, if you take your time to scan the waves from most British headlands, you stand a good chance of locating at least one of these sleek, elegant birds, bobbing on the surface or flying fast and low over the wave crests. This is the smallest diver species, and for most birdwatchers the one they are most likely to see. It nests inland, using even very small lochans, but can be seen on northern seas all year round as it often commutes to the coast to find food during the breeding season.

INTRODUCTION

This bird is named for the dark red throat-patch it shows in breeding plumage, making it the only non-monochrome diver, but unless you are lucky enough to see one closeup and in good light, the throat can look black. It has a distinctive head profile, with a smaller, slimmer and slightly upturned bill compared with that of the Black-throated Diver. In breeding plumage its upperparts are less boldly patterned than those of the other divers – its species name, *stellata*, means 'starry' and describes the fine white speckling on its upperside. In winter plumage the prominent eye (because the pale cheek extends further up than on other divers) is a good identification feature. The bird sits low on the water, and in flight looks very slim, with a long, thin neck and large feet on long legs (in fact its legs are not particularly long but are set so far back on the body that they project well beyond the tail when it flies).

The crimson throat-patch is exhibited in a neck-stretching courtship display.

This well-grown chick is fast losing its down layer, to reveal grey juvenile plumage.

DISTRIBUTION, POPULATION AND HABITAT

Red-throated Divers breed in northern and north-west Scotland, and there are strong populations on the major island groups, where the birds benefit from the absence of some mammalian predators. There is also a very small population (about ten pairs) in north-west Northern Ireland – these last birds represent the most southerly outpost in the entire breeding range. There are 1,000–1,600 breeding pairs altogether in the UK, swelling to around 17,000 individuals in winter. Most of the extra birds are Scandinavian breeders, but they could potentially come from further afield, because the species breeds across the whole of the Arctic, including North America, where it is known as the Red-throated Loon. In winter it moves south as far as Portugal in Europe, and Florida and Canada in North America. Its world population is somewhere around 200,000–590,000 individuals. The overall population trend is of decline, including in Britain. However, this has followed an increase throughout most of the 20th century, once the bird was given its fully protected status and persecution became infrequent.

The Red-throated Diver breeds in rugged, open and boggy uplands, around shallow lochans and sometimes also larger lochs. Often the water body on which it nests is some distance from suitable fishing grounds, in which cases it travels to the nearest coastline or to larger lochs to find fish. In winter it can be found all around the coast, although sheltered bays with sandy bottoms seem to offer it the best feeding opportunities. It is the least likely of the three regular diver species in Britain to visit reservoirs and gravel pits in the winter months.

Adults with chicks to feed will often fly to nearby lakes to catch fish, which they carefully carry back 'home'.

BEHAVIOUR AND DIET

This bird has a typical diver stance on the water, with a low profile and long, level back. It is much more at ease on the water than on land, and in flight is quite laboured, needing a take-off run and hard flapping to get airborne and stay in the air. It is, however, more aerially proficient than the larger divers, and unlike them is able to take off from the land.

This is not a particularly gregarious bird, usually feeding alone, but a productive area of sea may contain many Red-throated Divers, and flocks on the move are sometimes seen, especially ahead of bad weather. When not actively feeding it rests on sheltered water, spending much time preening and re-waterproofing its plumage.

A foot-propelled diver, this species submerges neatly with a slight jump and a strong backwards kick. Under water, it swims strongly and with agility, sometimes using its wings to facilitate a quick turn when chasing prey. It dives to about 10m down in a typical fishing foray, and usually stays under for between 30 seconds and a minute, but sometimes double that. Captured fish are brought to the surface to be eaten, or carried to the nest-site if the bird is a parent with dependent chicks to feed. The diet consists mainly of fish such as sprats, sandeels, herring and small cod. When hunting inland the birds take salmon, Brown Trout and perch. Non-fish prey such as crustaceans and insect larvae occasionally supplement this diet.

A coordinated 'rush' along the surface in upright posture helps a courting pair test out each other's fitness.

BREEDING

Most birds return to their nesting grounds in April. This species usually pairs for life or at least several seasons, although the pair may not see each other at all during the winter. A period of rebonding courtship behaviour takes place between established pairs when they reunite, with more prolonged courtship between newly formed pairs. This involves various posturing, dashes along the surface with the head stretched forwards, and the pair flying together and calling. This is a territorial species and a typical breeding lochan supports a single pair of birds, although a larger loch may hold more. First breeding takes place at two or three years old.

The nest is built in thick cover at the edge of (or sometimes floating on) the water, and is a scrape or mud platform lined with heaped vegetation. The clutch usually consists of two eggs, occasionally one or three. The eggshells are drab greenish-brown with darker spots. Incubation begins as soon as the first egg is laid, so that this egg hatches ahead of subsequent ones. This head start gives the oldest chick a distinct survival advantage – it may also be aggressive to younger siblings if food is in short supply. The female undertakes most of the incubation but is sometimes relieved by her mate – the off-duty bird may loaf on the water nearby or fly away to feed elsewhere. Incubation takes 24–29 days – weather conditions may affect the duration. The chicks are downy and open-eyed when they hatch, and can walk and swim within a few hours of hatching. They do, however, depend on their parents to feed them, and this may continue even after the young birds can fly (at about seven weeks old) and the family has left the breeding grounds.

When they are very small the chicks are offered insect larvae and very small fish, but they quickly progress to taking quite sizeable fish, which are usually flown in one at a time from elsewhere by the adults. While one parent is away on a food-collecting trip the other guards the brood, and may perform distraction displays to draw away potential predators. The most dangerous predators to diver eggs and chicks are mammals, particularly the American Mink but also Red Foxes and in some areas Pine Martens. Gulls and skuas may also take eggs and chicks.

WEATHER FORECASTER?

Like other divers this species is quite vocal in the breeding season, with a repertoire of mournful wailing calls. Its Shetland nickname of 'raingoose' is derived from the belief that when it gives particularly drawn-out calls, rain is on the way. These calls inspired the North American name for the birds, 'loons', as they are quite familiar to North Americans as inland breeding birds. In Britain, most of us only encounter them in winter, when they are silent and occur around our coasts.

MOVEMENTS AND MIGRATION

Some British breeding birds are thought to winter locally but many move further south, into English and French coastal waters, and perhaps further (especially in cold winters). There have been a few recoveries in British waters of Red-throated Divers ringed in Greenland and Finland, and many from Sweden and Norway.

Because their legs are set back so far on their bodies, divers are awkward on land and only come ashore when absolutely necessary.

THE FUTURE

The Red-throated Diver was a target for estate managers with fishing interests for much of the 19th and early 20th centuries, but its numbers began to increase somewhat after it was given full legal protection under the Protection of Birds Act in 1954. It is a Schedule 1 species (as are all divers), with extra protection from disturbance at its breeding grounds. It is extremely sensitive to this, although the remote and inaccessible nature of its preferred nesting habitat affords it some protection from unintentional disturbance that may result from activities such as hiking and other recreational land use.

The species' breeding habitat is at risk from peat extraction, afforestation and drainage, and in some areas the introduced American Mink has had an impact on breeding success. Climate change is also likely to have an impact, and may be responsible for the decline of the species in Ireland. High spring rainfall may result in nests being lost because of flooding, a situation that can be mitigated by providing floating platforms for nesting, as has been done successfully for Black-throated Divers. At sea birds may suffer the consequences of oil spills, or become entangled in fishing nets. Algal blooms ('red tides') at sea can harm them by poisoning through ingestion, and by damaging the feather structure and reducing waterproofing, resulting in hypothermia.

Like many seabirds, the Red-throated Diver should benefit from the establishment of more marine Special Protection Areas, where there are restrictions placed on activities that may threaten birds. On land sensitive management of moorland areas is needed to protect it while it is breeding.

In winter, Red-throated Divers can be seen offshore practically everywhere around the British Isles.

Black-throated Diver *Gavia arctica*

This strikingly beautiful bird is a rare and vulnerable breeding species in Britain, although it has responded well to conservation measures. It is also rather rare in winter, is the scarcest of our three regularly wintering divers and can be difficult to find, rarely gathering in any numbers.

INTRODUCTION

With its sleek outline and silky-smooth, 'painted on' black-and-white patterns, in breeding plumage the Black-throated Diver is one of the gems of the British bird fauna. In winter it becomes much drabber (although it is still neatly marked), and much more difficult to separate from the other diver species. It is intermediate in size between the Red-throated and Great Northern Diver, so can be easily confused with either of these species. Subtleties of outline and posture are sometimes more useful for identification purposes than the rather insignificant differences in plumage.

DISTRIBUTION, POPULATION AND HABITAT

The 190–250 pairs of Black-throated Divers breeding in Britain are restricted to north-west Scotland, where they nest mainly around medium to large, undisturbed lochs with small islands. They may use different lochs for feeding and breeding. In winter around 560 birds are widely scattered offshore, the highest numbers in sea lochs on the west coast of Scotland. There are also good numbers off south-west England, but very few move into the Irish Sea. Additionally, a few birds turn up each year on inland reservoirs. The species' global distribution is extensive, ranging across Scandinavia and Russia to northern North America (where it is known as the Arctic Loon). Its total population is somewhere between 280,000 and 1.5 million birds, although it is declining.

Winter-plumaged divers are difficult to identify. The Black-necked is in between the Red-throated and Great Northern in size and bulk.

BEHAVIOUR AND DIET

Awkward on land and laboured in flight, the Black-throated Diver spends most of its time loafing on the water, sleeping and frequently rolling onto its side to preen its belly plumage, or stretching out a foot behind its tail. It sits low on the water and holds its head level. When feeding it makes a neat dive with no obvious 'jump', and swims under water for more than 40 seconds, kicking strongly along with its large webbed feet. Caught fish are brought to the surface to eat.

The prey species favoured by the Black-throated Diver include Brown Trout and minnows on the breeding lochs, and sandeels, herring and sprats at sea. Crustaceans are sometimes taken, and on the lochs large insect larvae make suitable offerings for the chicks.

BREEDING

This strictly protected species is a very rare breeding bird in Britain, and very susceptible to disturbance when nesting.

Adults arrive on their breeding grounds in pairs in late spring. Courtship displays involving wailing calls, bill dipping and other ritualised movements are only likely to be seen in newly formed pairs. As this species forms long-lasting bonds, most couples will already be bonded and familiar with each other so there is little ceremony before they proceed to nest-building, although the arrival of a new diver triggers territorial displays by the resident pair. Both sexes work together on piling up vegetation to form a very basic nest, often on an island shore. The female lays two eggs and the pair shares incubation duties until the eggs hatch after 29 days.

The chicks take to the water within the first 24 hours of their lives and are tended by both parents, which feed small fish and insect larvae to them bill to bill; the chicks may also ride on the adults' backs. Sometimes a parent travels to a different loch or to the sea to fetch food for the chicks. The adults fiercely defend the chicks from potential predators and even chase off harmless birds such as Common Scoters. The chicks are also aggressive to each other. They are able to fly at 60 days old, whereupon the family leaves the loch and flies to the wintering grounds.

MOVEMENTS AND MIGRATION

Birds that breed in Britain are thought to remain in British waters over winter. The additional birds that arrive in winter are probably Scandinavian breeders, but the rarity of the species is such that there is little information available from ringing recoveries to shed light on its migratory habits.

THE FUTURE

This diver's few breeding sites are at high risk of disturbance – even the presence of small numbers of visiting birdwatchers can have an impact on nesting success. Changes in the balance of fish species as a result of management for angling interests can also render lochs unsuitable for breeding. Inclement spring weather can destroy active nests through flooding – this can be averted by providing floating platforms, a strategy that has been used successfully on Loch Maree in the Northwest Highlands, a key breeding site for the species. At sea it may encounter hazards such as oil pollution, although it is perhaps less vulnerable to sizeable losses in this way because of its rather scattered winter distribution.

Due to this species' habitat requirements it will never be a common breeding bird in Britain, but many apparently suitable lochs are unoccupied. The RSPB has carried out work on habitat requirements and has identified several ways in which loch management can be improved to better suit the Black-throated Diver's needs; it advises fisheries and other landowners accordingly.

Great Crested Grebe *Podiceps cristatus*

A lovely and distinctive water bird, resplendent in its ornamental breeding plumage, the Great Crested Grebe is a freshwater bird rather than a seabird in most birdwatchers' experience. It is best known for the wonderful courtship display that it performs on its breeding sites in spring, and is a common sight all year round on inland lakes, rivers and even parkland ponds. However, significant numbers also spend their winter pursuing a very different and truly oceanic lifestyle. The soft, dense body plumage ('grebe fur') of this grebe was once much in demand for the fashion trade, which led to a very severe decline in its numbers. Concern over this led to the first wide-scale bird protection laws to be passed in Britain in the late 19th century.

INTRODUCTION

This is the largest grebe species in the British Isles and one of two that are common and widespread, the other being the Little Grebe. It is a graceful, long-necked bird with a pale underside and darker upperparts, a straight, dagger-like pink bill and no tail to speak of. Its feet, like those of all grebes, are not webbed, but the three forward-facing toes have broad, fleshy lobes. Adults in breeding plumage are unmistakable, sporting a dark crest and an orange, dark-edged face ruff. These ornamentations are displayed to their best advantage during courtship. Young birds and adults in winter plumage are drabber, lack the elongated head plumes, and could be confused with other grebes. In flight the bird is diver-like, looking elongated at both ends, and shows white on both the leading edge of the wing and the secondaries.

Elegant and slender, with its ornamented head in summer, the Great Crested Grebe stands out amongst other common freshwater swimming birds.

Any fish that is small enough to swallow is fair game for this skilful piscivore.

DISTRIBUTION, POPULATION AND HABITAT

This grebe's breeding population is widely distributed throughout England, Wales and southern Scotland, and in Ireland. It breeds primarily on lowland fresh water with good fish populations and lush vegetation at the edges, especially slow-flowing rivers and good-sized lakes. It is quite willing to live and nest in large parkland lakes in cities, where it can become very approachable. There are about 12,000 adults in total in the British Isles during the breeding season, with there having been a gradual increase in recent years; in winter more birds arrive, bringing the total to more than 23,000, many of them spending the duration offshore.

Beyond Britain the species is distributed across Europe and throughout Central Asia, with many of these birds wintering further south. It has smaller and more scattered breeding populations in Africa (subspecies *infuscatus*), and Australia and New Zealand (subspecies *australis*). Its world population is in the region of 920,000–1.4 million birds.

Good breeding habitat is clear and not too shallow fresh water with high populations of fish and other aquatic life, combined with well-vegetated shores and islands. In winter Great Crested Grebes may flock on larger and more open, less vegetated fresh waters such as concrete-edged reservoirs, as well as offshore in reasonably sheltered seas.

BEHAVIOUR AND DIET

Great Crested Grebes rarely set foot on dry land except when on their nests. The setting of their legs at the rear of the body makes them very awkward on land. Their flight is laboured at first, requiring a long, pattering run along the water's surface, but once in the air they can cover long distances quite comfortably. However, they are most relaxed on the water and spend much time sleeping on the surface with their necks arched backwards so that they can rest their heads on their backs. In this stance the white breast is very prominent, and the birds sit rather high in the water.

Many adult Great Crested Grebes stay on their territory with their mate all year around, but young birds can be highly gregarious. At sea and on large reservoirs they assemble in rafts of dozens, sometimes hundreds of birds. These gatherings offer them the opportunity to meet partners for when they are ready to breed at two years of age.

A Great Crested Grebe's swimming posture is lower when it is actively feeding than at other times. It dives frequently, making a neat little jump, then powering down with strong kicks. Under water it is fast and agile as it chases its prey. Town lakes such as those in the London Royal Parks can offer wonderful opportunities to watch hunting grebes as they swim under water. After a successful dive a grebe brings its catch to the surface to swallow it.

This grebe feeds primarily on fish, taking all manner of small and medium-sized freshwater species. It also takes frogs and newts, and large insect larvae. At sea it feeds on fish such as herring and sprats.

Grebes are tender and devoted parents, often carrying their chicks on their backs to keep them safe from predatory fish.

BREEDING

New pairs form throughout the winter months, while many established pairs remain together through the winter. In all cases courtship behaviour reaches its peak when the birds are on their breeding grounds in early spring, although brief displays between a pair may be observed at almost any time of year, and may serve to reinforce an existing bond.

The nest is a heap of water vegetation and is often semi-floating, but anchored to an island. This gives it some protection against flooding (although many nests are still lost when water levels rise rapidly after heavy rain). The birds add material to the nest throughout incubation, to compensate for any bits that come adrift through the natural movement of the surrounding water. The clutch comprises two to six eggs, and the 28-day incubation period begins with the arrival of the first egg, so in a large brood there is a significant age gap between the oldest and youngest chicks. The eggs are whitish initially, but soon become stained with brown and green marks because of the parents' habit of covering them with nesting material before they leave the nest.

The chicks are active and mobile soon after hatching, and follow their parents or ride on their backs. Catching a lift affords them protection from Pike, which are key predators of all kinds of waterfowl chicks. Whether swimming or piggy-backing, the young grebes squeak incessantly for food, and one adult dives repeatedly to find food for them while the other minds the brood. Besides small fish, the parents also sometimes offer them soft feathers, which the chicks swallow – it is thought this helps to protect their internal organs from damage that could be caused by hard fish spines. Like all grebe chicks they have a stripy pattern when downy, and traces of this are still evident in juvenile plumage.

Young Great Crested Grebes can fly at about 11 weeks of age, and soon afterwards many move considerable distances, including to the sea, to begin their independent lives, while their parents embark on a second brood.

STRICTLY GREBE DANCING

The courtship in its full glory is rarely observed. A common element, often used as a greeting when a pair reunites after spending time feeding in different areas, is mutual head shaking. The two birds face each other with ruffs puffed out and crests raised, and take turns to rapidly shake their heads. The spectacular 'weed dance' is rarer, and involves both birds diving when apart, swimming together under water at high speed, then surfacing breast to breast and standing up in a vertical posture, each bird holding a billful of water vegetation. Another interesting posture, adopted by one bird at a time, is the 'cat display', whereby the bird dips its head and raises and half-spreads its wings to show off their white markings. Because grebes take equal roles in parenting, their roles within their display are also equal, and some parts of the display (such as the weed dance) are ritualised demonstrations of behaviours they will later use when rearing young.

A pair of Great Crested Grebes complete their courtship with the stunning 'weed dance'.

Most nests are at least partly fixed to land, making them vulnerable to flooding.

MOVEMENTS AND MIGRATION

British Great Crested Grebes – even young birds dispersing for the first time – are not thought to migrate very long distances in general. The extra birds that arrive here in winter also appear to have travelled relatively short distances. British-ringed birds have been found in France and Germany, while there have been recoveries in Britain of birds ringed in the Netherlands and Denmark. The oldest ringed British bird was close to 12 years old, but the recovery in Russia of a 19-year-old individual indicates that this can be quite a long-lived species.

A young Great Crested Grebe, still showing remnants of head stripes. Many youngsters spend their first winter living in flocks offshore.

THE FUTURE

This species' recovery as a breeding species in the British Isles has been remarkable. It was close to extinction in Victorian times because of hunting pressure, but is now a common, very familiar and much-loved species, with a population that is still growing (an 11 per cent increase was recorded in 1995–2011). It has benefited greatly from the many gravel workings around the country that were allowed to flood and slowly develop into lakes with rich natural flora and fauna, and more recently from relatively green urban developments that incorporate landscaped lakes. It has a fairly high tolerance to human presence and can become quite habituated to people provided disturbance is not severe. This helps to offset habitat degradation elsewhere, for example on rivers where increased boat traffic can cause quite severe disturbance and damage marginal vegetation, reducing available nesting habitat.

At sea Great Crested Grebes are vulnerable to the usual problems that face species which spend their time swimming on the surface, in particular oil spills and trapping in nets. Redistribution of favoured fish-feeding grounds because of climate change may also have a negative impact, but overall the present picture for this beautiful bird is quite a rosy one.

Slavonian Grebe
Podiceps auritus

This is the most marine of all of our grebe species, and it is not by any means common. Its British breeding population is tiny, and although numbers increase in winter it is still quite a scarce bird. It is the most colourful of our grebes and arguably the most beautiful when in breeding plumage; it remains a dapper and attractive little bird in its monochrome winter plumage.

INTRODUCTION

Of the five regularly occurring grebes in Britain, this is in the middle in size terms, although it is just half the size of the Great Crested Grebe. It is about the same length as a Moorhen but looks much slighter. In breeding plumage it is burgundy below and black above with a black neck-ruff, and broad golden 'eyebrows' extending into a shaggy crest, hence the American name Horned Grebe. (Its English name references Slavonia, a historical region of Croatia.) In winter it is neatly marked blackish above and white below, with the only hint of colour being provided by its red eyes. It has a small, straight bill, lacking the upwards tilt of the bill of the otherwise very similar Black-necked Grebe.

DISTRIBUTION, POPULATION AND HABITAT

About 30 pairs of Slavonian Grebes breed in Britain, all in the Scottish Highlands. RSPB Loch Ruthven is a key site; other breeding sites are usually kept secret to protect the birds from disturbance. In winter some 1,100 birds are present, although there are more in colder winters. Many of these spend the whole winter offshore, but a few stay on (usually coastal) inland freshwater lakes and reservoirs. The species' distribution at sea has a northerly bias, with favoured sites including Orkney, the southern

One of the smallest swimming seabirds, the Slavonian Grebe is nevertheless a consummate predator.

Firth of Forth and the Western Isles. Further south good sites include Poole Harbour in Dorset and the Blackwater Estuary in Essex.

The bird's world range is extensive, with a breeding distribution spanning Iceland, northern Europe, especially Scandinavia, central Russia, Alaska, Canada and the northern USA, and a more southerly winter range that reaches the Black and Caspian Seas, and in North America the coasts off Texas and California. Its world population is somewhere around 140,000–1.1 million birds – much of its habitat is inaccessible so accurate surveying is difficult.

The Slavonian Grebe nests around well-vegetated inland lakes, often in loose colonies. In winter most birds are found in sheltered bays and estuaries, sometimes in small parties. In Britain inland winterers are usually alone.

BEHAVIOUR AND DIET

Like other grebes this species spends most of its time swimming and diving, and is a rather reluctant flyer and a very reluctant walker. When feeding it makes frequent long dives (averaging 30 seconds), and you need to be quick to catch a look at it during its brief moments on the surface. It feeds on small fish and other small swimming organisms – in the breeding season aquatic insects make up the main part of its diet.

BREEDING

Slavonian Grebes pair up in the winter months; birds that have bred before are not especially likely to have the same mate in successive seasons. Courtship behaviour may be observed on the breeding grounds, and consists of various ritualised displays including a variant of the 'weed dance' (see also page 44), in which the pair swims in parallel, each bird carrying some water weed. The nest is a floating platform of weed, on which the female lays up to five eggs. Both parents share incubation (lasting 24 days) and chick-care duties, which include ferrying the youngsters about on their backs.

The young grebes can swim straight after hatching, but are not able to hunt for themselves for some 45 days. The parents feed them on water insects and small fish, and also give them feathers to swallow (for explanation of this behaviour, see page 44).

In its summer finery, this is our most colourful grebe species.

MOVEMENTS AND MIGRATION

There are few ringing recoveries involving this rare bird, but there are records of a Russian-ringed bird found in Yorkshire, and of a bird ringed as an adult in Scotland found in Iceland. This suggests that our wintering population includes some quite long-distance migrants.

THE FUTURE

This bird has a precarious toehold as a British breeding species. From a high of 70 pairs in the 1990s, numbers have fallen considerably and the reasons for the decline are not clear. Issues that can affect breeding success include stocking lakes with Brown Trout, which reduces the amount of insects available to the grebes, and disturbance. Like other very rare species, the Slavonian Grebe is also targeted by illegal egg collectors – nesting sites therefore need to be kept secret, or publicised but made inaccessible to those with nefarious intentions.

Some of the Slavonian Grebe's preferred offshore wintering sites fall within existing and proposed Marine Special Protection Areas, which should help to safeguard the winter population, although these birds are so few in number and so scattered in their distribution that it is difficult to apply much meaningful protection to them.

Other divers and grebes

The **Great Northern Diver** ***Gavia immer*** is the largest diver likely to be seen in British waters. It does not breed here – its breeding distribution covers Alaska, Canada, the northern USA and Greenland, and there is a small population in Iceland. However, a non-breeding bird may be encountered around the coast in summer, especially in the bays and sea lochs of north-west Scotland. There is a single confirmed breeding record from Wester Ross, in 1970, as well as records of hybrid Great Northern and Black-throated Diver pairs producing chicks on a handful of occasions. The Arthur Ransome children's novel *Great Northern?*, the last book in the *Swallows and Amazons* series, describes the discovery of a breeding pair of Great Northern Divers in the Western Isles, and the attempts of the young characters to protect the birds and their nest from egg collectors.

Around 2,600 Great Northern Divers arrive here in winter and they may be found anywhere around the British Isles, although again they are most likely to occur in good numbers in northern and western Scotland, with Cornwall being another favoured area. The sea lochs and channels in between islands in Orkney, Shetland and the Hebrides are particularly likely to support several Great Northern Divers. Inland records are relatively frequent as well. There is very little ringing data available for the species, but as only a few hundred pairs breed in Iceland it is likely that many of the birds that make up our wintering population originate in Greenland. Great Northern Divers also winter further south, reaching the coast of north-west Africa.

The massive bill of the Great Northern Diver helps distinguish it from its relatives.

This species is a large, stout, heavy-billed bird, with standard dull grey tones in winter but smartly marked black and white in breeding plumage. In North America, where it is abundant, it is known as the Common Loon and is well known for its unearthly wailing calls, which are given by birds on their breeding lakes. In Britain the birds that adopt a reservoir or gravel pit as their temporary winter home can often be very confiding, and allow prolonged, close observation. Offshore, however, they frequently feed further out than the smaller diver species, up to 10km, reflecting their greater diving ability and need for larger prey. As a result our estimates of wintering numbers may be too low. On the other hand, the species is declining on its breeding grounds.

Great Northern Divers are likely to be seen singly around the British Isles. They regularly make dives lasting a minute or longer, and prey on medium-sized fish, especially herring and large sandeels. They are heavy and ponderous in flight, with slower wingbeats than the smaller species. Around Britain they are very unlikely to be seen on land.

A rarity that draws much attention from birdwatchers, the White-billed Diver visits British waters in tiny numbers.

Two other divers have been recorded in British waters, although both are very rare. The **White-billed Diver** ***Gavia adamsii*** is similar in size and general appearance to the Great Northern Diver. It breeds in Arctic Russia, Alaska and Canada, and moves south along the coastlines for the winter. Some individuals travel prodigious distances – the species has been recorded off Mexico and Spain, and in most winters one or two birds are found offshore around the British Isles. Its American name is Yellow-billed Loon, although really the bill is neither white nor yellow, but pale horn. Its light colour is noticeable over some distance, as is its shape – longer and more angular than that of the dark-billed Great Northern Diver.

The other diver species is the **Pacific Diver** ***Gavia pacifica***. This bird is the eastern Russian and North American equivalent of the Black-throated Diver, from which it was split in 2007 – the same year in which no fewer than three individuals were identified in Britain, the first records for Europe. The species is very similar to the Black-throated Diver and identification is a significant challenge. One of the 2007 birds, an adult in Cornwall that visited for several successive winters, was seen and photographed alongside Black-throated Divers, allowing for some useful side-by-side comparisons. The two are distinguished by the Pacific's slightly smaller size and several subtle plumage differences, although with distant sea views it is not always possible to make a certain identification. With interest growing since the species was split from the Black-throated Diver, and British birders' awareness of Pacific Divers and how to separate them from Black-throated Divers, it seems likely that there will be many more records of the Pacific Diver in years to come.

The **Red-necked Grebe** ***Podiceps grisegena*** only occasionally attempts to breed in Britain, but there are regularly a few dozen here in winter, the majority of them offshore. This is the rarest of the five British grebe species at all times of the year. Globally it is very widespread, occurring further east in Europe and across Russia, north-east China, Japan and western North America.

This grebe is a little smaller than the Great Crested Grebe, but usually shows a similar graceful, long-necked and long-billed outline (although it can look hunched and compact in some postures). It sits higher in the water than the Great Crested Grebe and is quicker in its movements and a more active and energetic feeder, making an obvious jump when initiating its dive. It is a colourful bird when in breeding plumage, with a lovely rufous neck and breast, pale cheeks, and a dark crown and crest at the nape. The yellow bill-base is also striking, including in winter when the bird is otherwise rather drab. In flight it shows white on the leading and trailing edge of the wing, a pattern like that of the Great Crested Grebe.

The scarcest of our regularly occurring grebes, the Red-necked Grebe looks like a smaller, drabber Great Crested in winter.

The Red-necked Grebe has never been an established breeding species in Britain. A few birds oversummer here each year, and their locations are kept secret to protect them from egg collectors, but breeding attempts are infrequent and most oversummering birds have been singletons. The first known successful breeding was in Scotland in 2001. Although there is plenty of suitable nesting habitat (richly vegetated pools, lakes and slow rivers) for it here, the chances of it breeding regularly in the future are low as its population on the mainland is declining.

Most wintering Red-necked Grebes are found off the east or south-east coasts, where they favour sheltered bays and estuaries. There are usually fewer than 100 here, but a cold snap further north-east can drive larger numbers our way. A few winter inland, on large reservoirs or lakes, and both at sea and inland they may team up with winter flocks of Great Crested Grebes, unlike the smaller rare grebes which tend to go it alone.

Black-necked Grebes usually winter inland but will also use very sheltered areas offshore.

The smallest of the three rarer grebes is the **Black-necked Grebe *Podiceps nigricollis***. This bird is also a more southerly species than either the Slavonian or Red-necked Grebe. The first breeding record in the British Isles was in 1904. It is still a scarce breeding bird, with up to 50 pairs nesting each year (often fewer), and between 100 and 200 spending the non-breeding months in Britain. Numbers peak in autumn as birds heading further south make a stop-off here. The Black-necked Grebe is less likely to be seen at sea than the larger species; most wintering birds are found on reservoirs or gravel pits. However, a few winter in sheltered estuary mouths or bays, most of them on the south or south-west coasts.

The breeding-plumaged Black-necked Grebe is very attractive, with mostly black upperparts, rich rufous flanks and flared-out golden tufts of fine feathers on its upper cheeks, inspiring its American name, Eared Grebe. In winter it becomes blackish above and white below, and in this plumage it may be confused with the Slavonian Grebe. The two species can be told apart by their different head shapes – the Slavonian is straight-billed and rather flat-crowned, while the Black-necked has a tip-tilted bill and a rounder head with a noticeably steep forehead.

Although rare in Britain, the Black-necked Grebe has a large global population of 3.9–4.2 million birds, and occurs on all continents except Australasia and Antarctica. Some more northerly populations are migratory, including British breeders which in winter leave the well-vegetated lakes where they nest. The species often breeds in loose colonies, but these can be quite transient, with areas used for a year or two then abandoned. One site in Ireland was colonised dramatically in 1930 with 250 pairs, but numbers dwindled in successive years and there were none after 1959. The birds need lakes that support a good population of small fish and aquatic insects, with enough safe nesting sites for several pairs. The presence of a Black-headed Gull colony is also helpful, as the grebes benefit from the gulls' vigilance against predators.

Sites that hold nesting Black-necked Grebes are often not publicised unless access can be closely managed, because of the species' rarity and attractiveness to egg collectors. The bird is easiest to see in winter, and large coastal lakes at sites in the south-east and south-west regularly hold one or two individuals for several weeks. It is uncommon offshore, but also easy to overlook because of its diminutive size and frequent long dives. Calm bays offer the best chance of spotting one.

The **Little Grebe *Tachybaptus ruficollis*** is the most common grebe species in the British Isles (although the Great Crested is a close second). It is a familiar waterbird, living and breeding successfully on quite small lakes and ditches, and the loud, whinnying call of territory-holding birds is a common sound in well-vegetated wetlands throughout the British Isles. Tiny and round bodied, this grebe feeds on very small fish, amphibians and invertebrates, and is not well equipped to cope with rougher water conditions. It is therefore not often seen at sea, but may venture into very sheltered bays and river mouths, especially when there is a big freeze on the land. It is buff and brown in its winter plumage, so is easily distinguished from the other two small grebe species, which are black and white in winter.

The tiny Little Grebe is a very common breeding bird on all kinds of fresh water, but is seldom seen at sea. As in other grebes, the offspring may piggyback for safety.

Tubenoses

Many people's idea of a classic seabird would be a member of the order Procellariiformes, even if that name was not at all familiar to them. These birds, nicknamed 'tubenoses' by birders, are pelagic wanderers, spending months or years roaming the oceans. Most have long wings and are adept at using the uplift between wave crests to soar and glide effortlessly over huge distances. The order includes albatrosses, petrels, storm petrels, shearwaters and prions. Only four species breed in the British Isles - the group's centre of diversity is in the southern hemisphere - but more than ten others have been recorded as passers-by, skimming our coastline on their epic journeys.

Tubenoses have several interesting adaptations, including a stomach sac that produces a fatty oil – this is fed to the chicks but also spat at potential predators to drive them away. The bill is formed from seven distinct horny plates, and the nostrils are enclosed within tubes. The tubes may produce an obvious separate 'bump' on top of the bill near the base (as seen in storm petrels, for example), or be more streamlined (as in albatrosses). This structure enhances the birds' ability to locate the sources of odours, and ties in with their enlarged olfactory bulbs – brain structures involved in the sense of smell. While many tubenoses also dive to catch living prey (albeit near the surface), they are also scavengers and congregate in large numbers around a rich (and richly smelly) whale corpse, for instance. They also follow fishing boats and can be attracted by pouring 'chum' (mixed-up fish guts and fish oil) into the sea, a trick used by many tour operators running pelagic seabird-watching boat trips.

Worldwide, many tubenose species are threatened by the introduction of non-native mammals to their breeding islands, and some have been wiped out altogether. These problems have affected British-breeding tubenoses too, although to a lesser extent.

Seeing the scarcer tubenoses from land is a challenge, as their occurrence is so unpredictable. When weather at sea is bad, with strong onshore winds, they pass closer to land, but this does not make for comfortable viewing conditions, especially as the best viewing points are exposed headlands. However, the rewards for committed seawatchers can be incredible. No less rewarding – but in a very different way – is visiting a breeding colony of European Storm-petrels or Manx Shearwaters at night to witness how these incredibly charismatic birds conduct their private lives.

By day and night, the Fulmar is most comfortable overflying deep sea.

Fulmar

Fulmarus glacialis

The Fulmar is the most common and widespread of our breeding tubenoses, and is also the only one that nests in open situations and comes to its nest by day. It is therefore the easiest member of the group to see and get to know. The Fulmar has also historically bucked the general seabird trend in the British Isles by greatly expanding its range and increasing in number over the last century, after first breeding in Shetland in 1878. It is now one of the most widespread breeding seabirds of all, and one of the few to nest on cliffs in south-east England. However, the trend has now been reversed, and the species has been in slow decline since the late 20th century.

INTRODUCTION

The Fulmar is the largest of our breeding tubenoses, being a little larger than a Common Gull, with long, narrow wings and a rather heavy-set, thick-necked body. It has a white head and underparts, and smoky-grey back and wings. This colour scheme makes it look superficially like a Herring or Common Gull, and as a result it is often dismissed or overlooked. However, if you spend a moment watching its stiff-winged flight, alternating long glides with bursts of hard flapping, quite unlike the leisurely, loose-wristed flight of a gull, its true nature will become clear. It also lacks the black wingtips of those gull species – the wingtips are slightly darker grey than the rest of the wing. On land it is waddling and awkward, and only really equipped to manoeuvre around the immediate vicinity of its cliffside nest. A close view reveals a very un-gull-like face, with large, frowning dark eyes and the typical complex tubenose bill, formed of several distinct plates, with tubular nostrils and a hooked tip. There is a darker colour morph, known as the 'blue' Fulmar, in which the entire plumage is mid-grey in tone. This form is rare in Britain, and most likely to be seen in the far north.

Fulmars often use their webbed feet to help control speed and direction when in flight.

This seabird tends to form loose colonies, each pair seeking out a sheltered hollow not too close to other nesting birds.

DISTRIBUTION, POPULATION AND HABITAT

Fulmars breed around most of the British and Irish coastlines, including around offshore islands, wherever there are cliffs of any height. The highest numbers are on the most consistently rugged north and west coasts, with some three-quarters of the British and Irish population found across Orkney, Shetland and the Western Isles. However, there are also strong populations on the chalk and sandstone cliffs of the south-east coast. Fulmars may be seen offshore at any point around the coast, at any time of year. In total there are a little over 500,000 pairs breeding in the UK and Ireland, which represents 13–19 per cent of the total breeding population in the North Atlantic. There are estimated to be 1.6–1.8 million birds around our coasts during the winter, comprising a mixture of local breeders and visitors from more northerly colonies both east and west of the British Isles.

The world breeding range is very extensive, covering coasts across the whole of the North Atlantic and North Pacific. Estimates of the total population range from 15 to 30 million individuals, and overall the population is increasing. The most northerly birds migrate south some distance in the winter, but remain well north of the Equator. The subspecies found in Britain is the nominate *F. g. glacialis*, and there are two others – *auduboni* and *rodgersi* – which breed further east.

Fulmars are colonial nesters but are not inclined to pack together like some other cliff-nesting birds. Some cliffs hold hundreds of pairs and others a dozen or fewer. Pairs are often very well spread out on a cliff-face, each couple occupying a sheltered ledge or crevice, often among clumps of grass or other vegetation, usually close to the top of the cliff. They use cliffs made up of any kind of rock, and are not overly deterred by people using the beaches below – for example, there is a large and well-known colony on the relatively low chalk and sandstone cliffs of Hunstanton, a busy seaside town in north Norfolk. A few colonies are found on inland rocky outcrops, for example at Carrol Rock in Sutherland, which is 7km from the coast. Fulmars also nest on seaside buildings, especially older structures with suitable cracks and crevices. When foraging they roam considerable distances offshore, and are attracted to fishing boats.

BEHAVIOUR AND DIET

From early winter Fulmars can be seen around their breeding colonies. Some are resting on their nesting ledges. Others use updraughts to circle around the cliffs, skimming past their nesting spots and often making several passes before either landing or turning and heading out to sea. On the wing they resemble gulls, but the flight style is distinctive – the parallel-edged wings are typically held very straight and rigid whether gliding or rapidly flapping, while gulls have more angled wings and a more leisurely flap. If you are close enough you may hear the birds on the cliff calling to those in flight, with a harsh, rasping cackle or chuckle.

A good food source attracts many Fulmars, but it is every bird for itself when dividing up the spoils.

Like most petrels Fulmars take most of their food from the sea's surface or just below it. Food may be picked up in flight, but they can also make shallow, rather clumsy plunge dives when attacking shoals of sandeels or other fish close to the surface. The hooked bill-tip helps them to get a secure grip on living prey. They are strong fliers and can glide over long distances; they are also comfortable resting on the sea. However, healthy Fulmars are very unlikely to be seen on land anywhere except at their nest-sites. They take more live prey in the breeding season and more carrion in winter, using their strong sense of smell to home in on floating dead fish and other sea animals. Food is swallowed straightaway rather than being carried.

Fulmars eat almost any kind of animal prey that is accessible to them. Sandeels and other small shoaling fish form an important part of the diet when breeding; they also eat zooplankton and various other small swimming sea animals. The proportion of fish to zooplankton taken has been shown to vary considerably between different breeding colonies. Any kind of carrion is also taken, from fish remains (offal) thrown from commercial fishing vessels, to whale and seal corpses.

A powerful sense of smell draws Fulmars and other tubenoses to washed-up sea mammal corpses.

BREEDING

Fulmars are very long-lived birds, most reaching their thirties and some surviving into their forties or beyond. They also form lasting, sometimes lifelong pair bonds – the chances of any given Fulmar pairing with the same mate of the previous year are about 90 per cent. Therefore a Fulmar partnership can endure for decades, even though the birds go their separate ways for some weeks at the end of each breeding season. When they return to their breeding grounds in winter, each bird heads for its previous year's nest-site, and once both members of a pair are back, they reunite with cackling and bowing displays, stretching their heads forwards and puffing out their throats.

Young birds visit the breeding grounds from the age of four or five, but it usually takes them at least another year or two before they can find and defend a suitable nesting site and attract a partner. Alternatively, a youngster may enter the breeding population by teaming up with an older bird that has a nest-site but whose mate has not returned. The average age of first breeding is nine years – younger first-time breeders are much less likely to be successful. Breeding success also improves with experience, with the improvement being most marked in younger birds.

Like most cliff nesters the Fulmar does not make an actual nest, but lays its single white egg directly onto rock or in a small scrape in soft earth on its ledge. The female incubates the egg on the first day, then departs for about seven days, leaving the male on duty. Thereafter they take shifts of three or four days each, until the egg hatches after 52 days. The parents then take turns to make lengthy foraging flights to collect food for the chick, the non-foraging parent staying with it at the nest for at least the first two weeks. The food that the parents deliver to the chick is regurgitated and partly digested. The chick is downy and open-eyed when hatching, and can beg loudly to be fed, but it is otherwise rather helpless. It cannot move any distance and is very vulnerable to predators. This is where the adult Fulmar's secret weapon (see above) comes into its own, discouraging gulls and skuas that cruise along the cliffs looking for a chance to snatch a chick. Growing chicks also acquire the ability to defend themselves by this means.

Bonded pairs 'duet' with toneless, cackling calls when reunited at the breeding site.

The chick grows steadily and by about six weeks old has well-developed wing-feathers and is heavier than its parents. At about this age the parents stop feeding the young bird, and it waits for another four or five days before making its maiden flight from the cliff and heading out to the sea to forage for itself. Its excess weight helps to sustain it through these early days of independence, as it learns where and how to find food. Mortality is likely to be higher in the first year than in subsequent years, although as the young birds live offshore for several years this is very difficult to measure. Among adults the annual survival rate is an impressive 97 per cent.

A SICKENING SMELL

When approached by a potential predator (including a human) on the nesting cliffs, Fulmars often stand their ground rather than take flight (especially when there are eggs or chicks to guard). They have a potent defence that is deployed if the intruder gets too close – they forcefully disgorge their stomach contents, which if aimed well leave the would-be predator covered in stinking fishy oil. This can be a serious problem for the victim. Gulls and other birds that have been heavily fouled by Fulmar oil may suffer similar effects to those incurred by birds affected by oil pollution at sea, losing their ability to keep warm and dry, and perhaps to fly. The bird's English and genus names are derived from this oil-spitting habit – a combination of the Old Norse words *full* meaning 'foul' and *mar* meaning 'gull'.

MOVEMENTS AND MIGRATION

This bird does not have set, predictable migration routes, but can comfortably survive over the open sea and therefore has considerable potential to travel. Young Fulmars spend the first four or five years of their lives at sea, and may roam considerable distances. British-bred birds wander south, east and west – there have been ringing recoveries in northern Scandinavia and Russia, Svalbard, Iceland, Greenland and north-east Canada, France and Spain; in one exceptional case an Irish-bred chick was recovered on the coast of Senegal in West Africa. There have also been British recoveries of birds ringed in Scandinavia, Iceland and Canada. Established breeding adults are less likely to wander – they leave their breeding grounds in late summer but many return as early as October.

THE FUTURE

Fulmars are thought to have benefited historically from human activities at sea. Commercial whaling in the North Atlantic became very intense towards the end of the 19th century, and whaling boats discarded a great deal of remnants from their catches. Fulmars probably prospered more than any other seabird species from this bounty. As the whaling industry slowed down at the turn of the century, so fishing trawlers took their place as a food source for Fulmars. More food available would have boosted the survival rates of young Fulmars, placing pressure on existing breeding sites and forcing the population to spread southwards. Some experts have also suggested that the population expansion could have been facilitated by climate change, coupled with a genetic shift that allowed the birds to better exploit warmer waters.

Fulmars seem to be quite adaptable in terms of the food types they can use. Different studies on food brought to chicks have shown that in a number of colonies live-caught fish is the main food, in some it is zooplankton, while in others fish offal is most important – and even within the same colony the diet breakdown can vary significantly from year to year. However, after decades of increase, some colonies have been shrinking significantly and rather rapidly – for example, numbers on Foula and Unst in Shetland fell by 55 per cent between the mid-1980s and the start of the 21st century. The reverse is apparent in Ireland, with some dramatic increases having been recorded over the same time period.

CROSSING THE LINE

A threat facing Fulmars comes from long-line fishing, whereby fishing vessels launch thousands of baited hooks on long lines. As surface pickers, Fulmars are attracted to the bait, become hooked and are dragged under and drowned. A study from Norway in 1997–1998 indicated that 20,000 Fulmars were being killed in this way per year.

Other tubenoses are also vulnerable to this hazard. Long-line fishing is heavily implicated in declines of albatross numbers in the southern hemisphere, and unlike the Fulmar, several albatross species are threatened with extinction. The RSPB's Save the Albatross campaign offers three key measures that make long-lining less dangerous to seabirds: weighting hooks so that they sink more quickly, using bird-scaring streamers around the ships and setting hooks at night when seabirds are not as likely to be foraging. Similar measures have been trialled in the Norwegian Sea and were found to be effective at reducing by-catch of Fulmars.

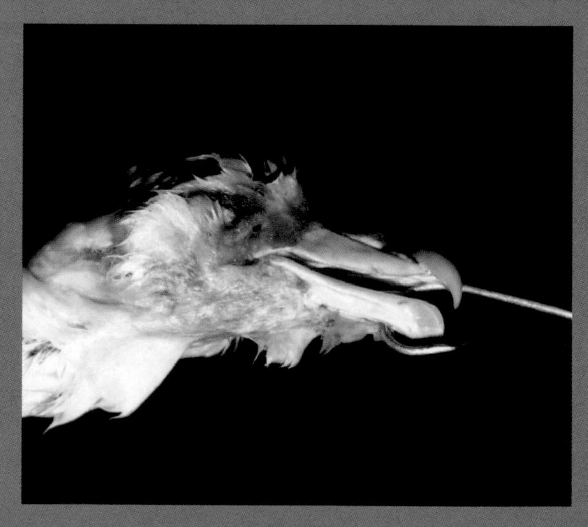

A baited fishing hook of any kind is potentially deadly for Fulmars and other birds that pick food from the sea surface.

Beware the Fulmar adopting a threat posture: if provoked further it may well eject its fishy, oily stomach contents.

This is the only tubenose that is easy to observe in most of the British Isles, and at present its population is stable.

Fulmars are threatened, as are all seabirds, by a general depletion of marine life. They are particularly likely to be affected by declines and redistributions of sandeels and similarly sized fish, and this may be the reason behind declines in populations dependent on the northern North Sea, where fish stocks have been dwindling severely since the 1970s and are only now showing some recovery with more prudent management. Due to reduced fish stocks there has also been less discarded offal available for Fulmars.

Fulmars are less susceptible to the effects of surface pollution than species that spend most of their time swimming, but can still be affected. The Fulmar was the seventh most numerous seabird picked up dead on beaches following the *Braer* oil spill in 1993. Examination of these birds' stomach contents also found that half of them had ingested plastic – although it is unclear how much this contributes to Fulmar mortality.

Overall, Fulmar numbers are currently fairly stable. Many colonies are probably at capacity with all or nearly all suitable nest-sites being used. It is also encouraging to note that some colonies are still growing, both in the British Isles and further afield, in Iceland, Norway and Canada. The steep declines in the Shetland population are cause for concern, and are likely to have the same root causes as declines in other seabird populations in the same area. However, the Fulmar's large and widespread population does give it strong potential to recolonise depleted areas should conditions become more favourable.

Manx Shearwater *Puffinus puffinus*

Tipping and tilting, brushing the crest of a wave with a dipping wing-tip, the Manx Shearwater rides the low-level oceanic air currents with consummate skill. Its 'shearing' glide, alternating with rapid beats of its extraordinarily long and narrow wings, is distinctive over any distance. On land its small size and clumsy gait render it extremely vulnerable to predators, so it nests in the safety of deep burrows and only comes ashore when darkness falls. Around 80 per cent of the world's Manx Shearwater population breeds in Britain and Ireland, making its conservation here a high priority.

INTRODUCTION

A fairly small, short-tailed and long-winged seabird, the Manx Shearwater has a long, thin bill with a slightly bulbous and hooked tip, and its front three toes are webbed. Its nostril tubes have a low profile and are only evident at very close range. Its plumage is sooty blackish-brown above and clean white below, with an area of smudgy-grey on the breast-sides where dark meets light. In flight it looks alternately black then white when 'shearing' with its slim wings held out stiff and straight. The plumage shows more white than that of other similar-sized (and rarer) shearwaters. The bird can look surprisingly auk-like in flapping flight, when it is harder to make out the wings' full length. Although silent at sea it is very noisy at its breeding colonies, making unsettling, almost demonic cackling and gasping noises. The Manx Shearwater is famously long-lived – it is perhaps Britain's longest-lived bird species, with at least one ringed bird reaching its fifties.

A storm-driven Manx Shearwater rests on a canal. Only very severe gales will force this seabird to take refuge inland.

As dusk approaches, Manx Shearwaters prepare to come ashore to their nest burrows.

DISTRIBUTION, POPULATION AND HABITAT

Manx Shearwaters have breeding colonies on the west side of Great Britain and both coasts of Ireland. Colonies are typically on islands, with particularly significant concentrations on the Pembrokeshire islands of Skomer and Skokholm, the inner Hebrides, and islands off County Kerry. Outside the breeding season birds may be seen offshore anywhere around the British Isles, especially in migration periods. The bulk of the population migrates to the seas around South America for the winter. The British and Irish breeding population has until recently been thought to total around 330,000 pairs, with the largest colonies (on Rum in the Inner Hebrides and on Skomer island) having more than 100,000 pairs each – amounting to 90 per cent of the UK's population. However, as surveying techniques improve, evidence is emerging that some colonies may be much larger than previously supposed. The remainder of the world's population breeds on Iceland, the Azores, Portugal, the Canary Islands and, as of the 1970s, on islets off the Atlantic coast of northern North America.

By nesting in burrows and small crevices, Manx Shearwaters gain some protection from predatory birds like skuas, but not from small mammals. Islands that are free of rats, mustelids and cats are thus most likely to hold good numbers of breeding Manx Shearwaters. The birds need soft earth in which to dig their burrows, and some may use holes in existing old Rabbit warrens, or crevices among boulders. Manx Shearwater guano can greatly enrich soil and leads to the establishment of a characteristic plant community.

RATTED OUT

The colony for which the species was named, on the Calf of Man (a 250-ha island off the Isle of Man) was wiped out when Brown Rats arrived via a wrecked boat in the 18th century. However, the island has since been recolonised, albeit in small numbers. A plan to eradicate the rats, initiated in 2012, should allow the shearwaters to return in good numbers. A similar scheme on Lundy Island, off Devon, was carried out in 2006 and resulted in a tenfold increase in the breeding shearwater population by 2013, to 3,000 pairs.

A feeding Fin Whale attracts swarms of Manx Shearwaters, ready to snap up prey scattered by the cetacean.

Manx Shearwaters can withstand heavy weather at sea, but may retreat into bays in very severe conditions and are occasionally 'wrecked' on inshore waters or land. One such event occurred in September 2011 when hundreds were blown ashore at Newgale in Wales. The RSPCA picked up nearly 500 grounded and exhausted Manx Shearwaters after the storm had passed, and about 80 per cent of these birds recovered well enough to be returned to the wild.

BEHAVIOUR AND DIET

This is a true seabird that is much more at home on or over the water than on land, and in its long lifetime it racks up a huge number of air miles, many of them thousands of kilometres from the nearest land. Flying low and close to the waves, it uses dynamic soaring to reduce the energy costs of flight, although it does not employ this technique as frequently or fully as some of its larger relatives. It often rests on the water's surface, sitting buoyantly with the wings well clear of the water, and needs to make a running start to take to the air again. It only comes ashore to breed, and is slow and ungainly on foot.

This seabird swims easily and buoyantly, but cannot take off from water without a running start.

Manx Shearwaters are most likely to feed at zones where the sea shelf falls away, resulting in upwellings that bring plankton near to the surface – the plankton in turn attracts fish, squid and other organisms on which the shearwaters prey. They feed principally by picking from the surface, but also by diving, either from low flight or while swimming. Under water they use their wings and feet for propulsion. The long, narrow wings look singularly ill adapted and even too fragile for this, but the birds can manage surprisingly deep dives, down to about 3m. They often travel and feed in flocks, which can be large, and even when breeding their foraging flights can cover long distances – satellite-tracking studies have recorded flights of 500km or more by an off-duty parent during incubation.

Analysis of the diet has shown that these birds feed primarily on fish and small cephalopods, with a smaller proportion of crustaceans. Some birds' stomach contents included a little plant material, although this may have been ingested accidentally. As well as taking live prey, Manx Shearwaters scavenge remains left by other predators, and follow fishing boats for scraps. Like other tubenoses they appear to have a well-developed sense of smell, and are attracted to chum.

BREEDING

Manx Shearwaters return to their breeding grounds at any time between late February and early April. Because they visit their colonies at night-time, they are able to recognise and locate their partners by sound – males tend to call for their mates from the ground, and females from the air. Established breeding birds nearly always pair with the same mate and use the same burrow year on year, although if they fail to breed successfully a 'divorce' and/or a burrow move sometimes occur in the following year. Young shearwaters in their first breeding season (five or six years old) seek to pair with older birds whose mates have not returned over the winter, as well as 'divorcees'.

The birds may visit the breeding grounds regularly over several weeks before egg laying commences. They spend this time renovating their burrows and building up their physical condition. The week or so immediately before laying is spent at sea, with the birds concentrating on taking in plenty of food to stock up their fat stores, ready for the demands of long incubation shifts, and in the case of the female to develop the single egg she will lay. The egg is large relative to the bird's size (15 per cent of body weight), and puts a considerable strain on the female's bodily resources.

The date of laying varies across the colony, but most of the females lay their white-shelled egg in the first half of May, and remain with it in the burrow for some hours. The male of each pair then takes the first long incubation shift, after which the birds alternate. Shifts last five days on average, but may be as short as two days or as long as 16 days; they tend to become shorter as the 50-day incubation period progresses. Close to the hatching date the eggs can survive being unattended for a day or more, as the burrows are well insulated and the ambient air temperature is usually high. Overall, about 80 per cent of eggs hatch.

The nesting burrow is small enough to keep out marauding gulls and skuas.

Manx Shearwater chicks are fluffy and open-eyed on hatching, but cannot move around very much and are very vulnerable to predators. Nesting burrows are not accessible to most predatory birds, but small mammals are another story, hence the much reduced breeding success on sites where rats are present. If this danger is not present the chicks have a high survival rate; at 90 per cent it exceeds the hatching rate. Some, however, succumb to a viral disease called puffinosis – this disease is usually uncommon, but there are occasionally epidemics that cause high mortality.

This tubenose has a delicate profile, its tubular nostrils barely breaking the outline of its slender bill.

Both parents feed the chick on regurgitated food brought on nightly visits, although sometimes the chick goes hungry for a night or two if the adults fail to find food – on average the adults are away for 1.65 days per foraging trip when feeding chicks, compared with 5.46 days during incubation. When the chick is about 60 days old, well fed and a third heavier than an adult Manx Shearwater, its parents stop coming to attend to it and its calls for more food go unanswered. It waits a further eight or nine days, sometimes venturing out of the burrow to flap and stretch its wings, before leaving for good at night to begin its new life as an independent seafarer.

The first flight is hazardous, as the chick's wings are still largely untested. Some chicks jump from the nearest cliff-edge, while others stand in the open with wings spread and try to lift off on the wind. It is essential that they clear the land before morning, when they would become easy prey for gulls, skuas and raptors on the ground. However, once flying at sea they are much safer and can quickly range far from land.

MOVEMENTS AND MIGRATION

These birds are true migrants and cover thousands of kilometres each year, travelling to the Patagonian shelf off Argentina, well south of the Equator. This journey, of some 10,000km, takes them across the Atlantic and down the coast of South America. Satellite-tracking studies indicate that the birds travel southwards down the west coast of Africa, then take the shortest route across to Brazil to continue down the east coast of South America. On their northwards return journey they take a less direct route, crossing over through the eastern Caribbean. The studies have also identified key feeding areas close to the final destinations, which are used by the birds for stopovers of about two weeks. These are south-east Brazil on the southwards flight, and areas in the middle of the western Atlantic, adjacent to France and Britain, on their northbound journeys.

Young Manx Shearwaters' movements are not well understood, but tracking technology should help to shed more light on them in years to come. Although the majority probably follow the same migratory routes as adults, they do not return to the breeding colonies for five or six years and may wander elsewhere in the interim. Ringed British youngsters have been recovered from places not on the regular migration route, including Greenland, Norway and Canada.

LOSING COUNT?

A count on Skomer in 2011, using playback but with more developed methodology than in previous studies, suggests there are now in excess of 300,000 pairs of Manx Shearwaters on the island, a staggering increase from the 101,000 pairs recorded for Seabird 2000 in 1998. This may be the result of a genuine population increase, or it could be entirely down to the improved counting method used and may not reflect a significant increase at all. Future surveys of other colonies should help clarify the situation, and over time build a picture of population trends.

Nearly full-grown chicks are at tremendous risk from predators like Lesser Black-backed Gulls if they emerge from their burrows by day.

THE FUTURE

The Manx Shearwater is a difficult species to survey because of its nocturnal habits and underground nesting, and counting breeding adults without causing serious disturbance is a challenge. It is sometimes possible to tell whether a burrow is in use by external signs of recent use, but some occupied burrows show no obvious signs of use. The least invasive but still accurate method is to use playback of recorded calls to see if incubating adults respond from within their burrows. This was used extensively in the Seabird 2000 project to give the first really accurate counts of Manx Shearwater abundance. These results, compared with previous counts using less sophisticated methods, suggest that the populations of some key colonies are stable or increasing – in some cases dramatically.

Manx Shearwaters have quite a diverse diet and are also long-range foragers, willing and able to cover hundreds of kilometres to find good food sources. They may therefore be less affected by changes in fish stocks and shoal distributions than other, less mobile British seabirds – although of course they could not escape the consequences of very serious degradation of marine ecosystems. Manx Shearwaters are at low risk from oil spills as they disperse widely and spend relatively little time swimming, although they have been found to ingest considerable volumes of plastic waste. The most serious threat they face on the breeding grounds is nest predation from introduced mammals, especially rats. Successful rat-eradication projects have been carried out on islands such as Lundy, and these have shown that Manx Shearwaters can rapidly recolonise suitable breeding habitat once rats are removed.

Other factors, such as very rainy spring weather causing mass flooding of burrows, and outbreaks of puffinosis, have the potential to dent Manx Shearwater numbers. Early autumn gales can sometimes 'wreck' large numbers of newly fledged young birds. The species' natural longevity, however, does give it a buffer against the occasional bad breeding season.

Because the British Isles hold the bulk of the world's breeding population, the Manx Shearwater is an important and iconic species and its continued conservation is a high priority. The main colonies are rightly recognised and safeguarded as precious components of our ecological heritage, and the designation of more Marine Special Protection Areas should also benefit the species.

European Storm-petrel *Hydrobates pelagicus*

This remarkable bird, barely larger than a sparrow, is in its element roaming over the open sea in the wildest weather, despite its tiny size and frail appearance. It is sometimes nicknamed 'Mother Carey's Chicken', after a mythical witch-like figure personifying dangerous seas. Great Britain and Ireland between them hold more than half of the world's breeding population, mainly on offshore islands where the petrels nest in large colonies. Seeing European Storm-petrels from land is difficult, as they usually feed some way from the shore. Even when closer in they are hard to spot as they are so small and tend to fly close to the water, lifting and dipping in between wave crests as they search for food. When the breeding birds come ashore at night, however, close and magical encounters are guaranteed as the air is filled with their fluttering forms and purring calls.

INTRODUCTION

The European Storm-petrel is mostly brownish-black, with a square white rump-patch, an obscure pale line along the centre of the upperwing and a broader, more prominent white line along the underside of the wing. It has a square-ended tail and a relatively large head with a steep forehead. The bill is small and looks low-set on the head, with a prominent 'tubenose', and long, feeble and slender legs that it often dangles while in its slow feeding flight. The habit of this and related species of pattering the feet on the water's surface gave rise to the name 'petrel', after St Peter, who was said to have walked on water. For its size this bird has large wings that are long, relatively broad and round ended, giving it low wing-loading. This allows it to employ a quite vigorous fluttering, bat-like flight style with much manoeuvring and relatively low energy costs, although it makes it ill suited to the dynamic gliding flight used by its larger and longer winged relatives. It swims comfortably, sitting buoyantly on the water.

Barely larger than a sparrow, the European Storm-petrel seems incredibly frail but easily copes with life far out at sea.

AN ICONIC HOME FOR SPECIAL BIRDS

On Mousa in Shetland stands the island group's finest and tallest broch – an ancient round drystone tower constructed in the Iron Age more than 2,000 years ago. The broch stands more than 13m high, and has a diameter of 15.2m at its base. The walls are up to 5m thick, and the bottom-most 4m of the interior are filled with a solid mass of stones. The original purpose of the broch is not known, although its shape, very sturdy construction and location on the coast suggests a defensive function. Today the outer shell of the broch is home to some of the island's 7,000 or so breeding pairs of European Storm-petrels, which use gaps between the stones as nesting tunnels. They also nest on boulder beaches on the island's shore. The RSPB manages this uninhabited island as a nature reserve, safeguarding the petrels and a wealth of other wildlife. Visitors can make day trips to Mousa by boat from Sandwick, to enjoy this extraordinary site.

DISTRIBUTION, POPULATION AND HABITAT

The European Storm-petrel is a highly localised breeding bird in the British Isles, and is almost completely restricted to small, mainly remote islands and islets. However, some of its colonies are very large. The main concentrations are on Scilly, the Pembrokeshire islands, Orkney, Shetland, the Western Isles and various small islands off the west coast of Ireland. The Seabird 2000 survey identified 95 colonies, only one of which was on the mainland. There are likely to be more colonies in existence, but surveying for this species has very significant practical difficulties. Some of the most important colonies, with several thousand breeding pairs, include those on Mousa in Shetland; the Treshnish Isles; Priest Island in the Summer Isles; Skokholm off Pembrokeshire; Great Skelling, Scariff, Inishtooskert and Inishvickillane, all off County Kerry; and Inishark off County Galway. In total there are an estimated 25,000 pairs in the UK and around 100,000 in the Republic of Ireland.

Beyond the British Isles the European Storm-petrel is also found in large numbers on the Faroe Islands and Iceland, and it has smaller outposts on small islands off France, Spain, the Canaries, Italy, Malta, Greece and Norway. Its wintering grounds are mainly off the coast of southern Africa. The world population is 1.29 million–1.53 million individuals.

The European Storm-petrel is very vulnerable to mammalian predators – especially rats – when nesting, so is only successful on sites that are free of these or inaccessible to them. The actual nesting habitat is anywhere that has an abundance of holes or hollows deep enough to keep the birds safe from marauding gulls, skuas and other predatory birds. In soft ground the birds may dig their own tunnels, or extend existing ones. Suitable pre-existing nesting holes include Rabbit burrows, gaps among piles of boulders and even crevices in buildings. On occasion a pair may share its burrow entrance (although not the actual nest chamber) with Puffins or Manx Shearwaters, although in the event of a dispute over ownership the petrels will be evicted (or even killed) by the larger species. They have also proved willing to use artificial tubular nesting boxes. Such a nestbox, if carefully designed, can provide a safer home than natural hollows and also allows easy access for monitoring, so it is a useful tool for scientists and conservation workers. When not breeding the European Storm-petrel is fully pelagic and may wander 200km a day over the sea as it searches for food.

Petrels are named for St Peter, as both the birds and the saint are known for 'walking on water'.

BEHAVIOUR AND DIET

The European Storm-petrel, like other tubenoses, is primarily a surface picker and can locate food over long distances using its sense of smell. Large numbers may gather at a suitable food source and it follows fishing boats; however, it also travels alone. The flight action is distinctive, with much flapping, although it also uses gliding flight on occasion. When actively feeding its rate of progress slows down dramatically as it patters on the water's surface with dangling legs and wings held raised. Although it 'walks on water' it cannot walk on land because its very slim legs will not support its weight. It can only shuffle, resting on its ankle joints, and is never seen on land away from its nesting sites unless it has come to grief in some way.

The European Storm-petrel's diet is mainly composed of zooplankton picked from the sea's surface. It occasionally plunges into the water in pursuit of larger prey, but it cannot easily dive beyond about a metre under water. This is enough, however, to allow it access to a range of other live prey types, which make up the bulk of its diet. It takes various small fish such as sprats and sandeels, and also crustaceans such as shrimps, and very small squid and jellyfish. It also eats floating carrion of all kinds, and is attracted to scraps thrown from fishing boats, as well as chum put out by birding tour boats.

Breeding European Storm-petrels travel long distances to find food, but tend to stay well offshore by day to avoid gulls and skuas, and feed closer inshore at night.

BREEDING

European Storm-petrels arrive at their colonies in mid-May, and birds that have bred before seek out their previous nest-sites and mates. Long-term fidelity to both nest-site and breeding partner is commonplace, although pairs that have failed in one year may not reunite the following year. Young birds seeking to breed for the first time may pair together if one can find a suitable unoccupied burrow, or may team up with an older 'divorced' or widowed bird. The sound of birds already on site, exchanging their excitable purring and chattering calls, attracts newcomers. Using a combination of recordings and artificial nestboxes, scientists have successfully encouraged various small petrel species to colonise new breeding sites, and recolonise historical sites following successful rat-eradication programmes.

The female European Storm-petrel lays a single egg inside the unlined nest chamber. The egg is white and, as is typical for tubenoses, proportionately very large, weighing almost 7g (relative to the adult's weight of about 28g). Both parents incubate the egg, taking shifts of about three days at a time while the off-duty parent forages at sea. If the off-duty bird stays away too long the other parent will need to leave the nest to forage for itself, leaving the egg alone. The incubation period is about 40 days, but can be up to ten days longer if the egg is left unattended for extended periods. The ability of the embryo to survive a certain degree of chilling is an adaptation for a species that may have to make extended foraging trips, although for most of the time there will be one parent on the egg.

The adults' visits to the breeding colony take place after dark, to reduce the chances of running into predatory birds. However, in some areas the local predators have adapted their own behaviour to

A crevice among boulders offers a safe place for a European Storm-petrel nest.

exploit the petrels. In the Mediterranean, Eleanora's Falcons are active well into dusk and catch petrels arriving at or leaving their nests. On Skomer, Little Owls (which are not native to the UK) are important predators of European Storm-petrels.

The newly hatched chick is clothed in light grey down and has open eyes. One parent remains with it while the other forages until the chick is about a week old – thereafter it is left alone throughout the day, and the parents visit at night to feed it. Its diet is the usual tubenose fare of regurgitated, partly digested fish and other prey, mixed in with the parent's stomach oil. Each nightly meal weighs about 1.3g. On this substantial and high-calorie diet, the chick builds prodigious fat stores, and by about 50 days old it weighs nearly half as much again as an adult. At this point the parents reduce the frequency of their visits, then stop them altogether, and most chicks leave their nests at around 62 days old, flying at night to avoid the attentions of predatory birds. They are independent from this point.

European Storm-petrels are extraordinarily long-lived for such small birds, with many living to their early teens and several ringed birds known to have reached their early thirties. Similarly sized land birds very rarely reach double figures. This long lifespan helps to offset the European Storm-petrel's very low breeding productivity, and explains why each pair invests so heavily each year in a single chick.

The Mousa birds arrive late in the evening and can be watched at close range as they climb to their nests among the stones.

MOVEMENTS AND MIGRATION

The European Storm-petrel is a long-range migrant, travelling many kilometres south to spend winter mainly off the coast of Namibia and South Africa. Birds begin to depart through September, and reach their wintering grounds in November. Ringing recoveries indicate that the migratory route tracks down the west side of Europe and Africa. Some of the Mediterranean breeders (which belong to a different subspecies – *H. p. melitensis*) do not migrate any distance but overwinter close to the breeding grounds.

Young birds do not revisit breeding colonies until they are about three years old, and may wander away from the mature birds' regular migration route. There have been recoveries of young British-bred European Storm-petrels from Sweden, Germany and, most surprisingly of all, Switzerland. No British-ringed birds have been recovered from the other side of the Atlantic, but there are regularly a few European Storm-petrels recorded on the US coast in autumn.

THE FUTURE

Accurately assessing the population of this seabird is immensely challenging, perhaps even more so than in the case of the Manx Shearwater, another nocturnal tunnel dweller. Using playbacks of calls is the most effective method to determine whether a burrow or crevice is occupied (if a bird is present it will answer a playback) and this was used extensively for the first time in Seabird 2000. Subsequent surveys will reveal the current population trend.

The biggest threat European Storm-petrels face is from predatory mammals introduced (accidentally or deliberately) to their nesting islands. If rats or cats become established on a European Storm-petrel island they wipe out the colony. It is therefore vital that these mammals are kept away from the islands occupied by European Storm-petrels, and if accidental introduction should occur, an aggressive eradication campaign is required. Elsewhere in the world petrel species closely related to the European Storm-petrel have been driven to extinction by introduced mammals. For example, cats were introduced to Guadalupe island off Mexico in the late 19th century; they bred freely and preyed heavily on the endemic Guadalupe Storm-petrel. The petrel, considered previously to be 'abundant', became extinct around 1912.

In most respects European Storm-petrel colonies are relatively easy to manage and protect. Disturbance by human activity is a potential problem, but most accessible colonies are well protected against this. Many others are on islands so remote and tiny that they are never visited by people, let alone other mammals. Dense colonies may suffer outbreaks of disease or parasite infestation, but these events are rare.

At sea European Storm-petrels are at relatively low risk from oil spills, but may over a long period of time ingest dangerous amounts of floating plastic or toxins. At present the signs are good that the population is stable, but it is important to protect their main feeding areas and work to maintain marine biodiversity in general.

TINY TRACKERS

Satellite-tracking technology as a tool to study wild birds' movements has advanced at a tremendous pace since its inception in the 1980s. The first bird to be tracked in this way was a Bald Eagle, a large and powerful bird that was able to carry what was then a rather heavy piece of equipment, a 170g transmitter mounted on a backpack. Since then the devices have become smaller and their efficiency and battery life have improved, but devices small and light enough to be placed on sparrow-sized birds have only recently been developed. The first satellite-tracking study on European Storm-petrels took place in 2011, when 34 European Storm-petrels breeding on Filfla, Malta, were fitted with tiny radio tags. The birds were monitored via light aircraft overflying the area, and resultant data was used to identify the marine areas most used by the birds when they were foraging. In the near future it should be possible to carry out tracking studies on European Storm-petrels as they migrate, and discover more about the route they take, stop-offs they make, and ways in which young pre-breeding age birds differ from breeding adults. This kind of data will help to establish where and how conservation efforts should be focused to protect our European Storm-petrels as they migrate.

Leach's Storm-petrel
Oceanodroma leucorhoa

Although many thousands of Leach's Storm-petrels breed in the British Isles, seeing them is a tall order for the average birdwatcher, because the vast majority of breeding birds live on St Kilda, a group of small islands far out to the north-west of the Western Isles. Mainland birdwatchers do stand a chance of seeing Leach's Storm-petrels when the birds are migrating south, especially when gales blow them inshore and sometimes even inland. Leach's Storm-petrel is one of our least-known breeding birds, but it has a huge global population and is one of the world's most abundant seabirds.

INTRODUCTION

At first glance Leach's Storm-petrel looks very similar to the European Storm-petrel, being small and dark with a white rump-patch. Leach's Storm-petrel is larger, closer to the size of a Starling than a sparrow, and the dark parts of its plumage are more grey-toned. Its tail is longer and slightly forked rather than square cut, its white rump-patch is U-shaped rather than square and has a narrow dark line down its centre, and the pale line across its upperwing is more prominent. Once familiar with the two species in the field, it becomes relatively easy to identify a Leach's Storm-petrel without the close view necessary to make out these differences, as the general impression of Leach's in flight is quite distinctive, lighter and more tern-like than that of the fluttering European Storm-petrel, with more gliding phases.

Although only a little larger than the European Storm-petrel, Leach's Storm-petrel is clearly longer-winged with a more dynamic flight.

Migrating Leach's Storm-petrels are sometimes forced onto beaches or even inland by heavy weather, but are usually quite able to get themselves back on track.

DISTRIBUTION, POPULATION AND HABITAT

About 94 per cent of the British and Irish population of Leach's Storm-petrels breed on St Kilda, more than half of them on Dùn, a 1.6km-long narrow island just south of Hirta (the largest island in the archipelago). Seabird 2000 found there to be some 27,700 pairs on Dùn and 45,400 on St Kilda altogether. However, a repeat survey on Dùn in 2006 found only 12,800 apparently occupied nests. There are also small colonies on several small islands in the Western Isles, on Foula and Gruney in Shetland, and on the Stags of Broadhaven in County Mayo. Seabird 2000 data indicated that the entire population in the British Isles is 48,000 pairs, but due to the declines on St Kilda the number could now be significantly lower. However, it is likely that other undiscovered colonies exist on similar remote tiny islands.

On some islands Leach's Storm-petrels breed alongside European Storm-petrels. However, many more islands hold European Storm-petrels but no Leach's. This may be because Leach's Storm-petrels need to breed within easy reach of waters where there are upwellings from very deep water by oceanic shelves, while European Storm-petrels are more adaptable. All known British and Irish Leach's colonies are no more than 67km from the continental shelf break.

Migrating Leach's Storm-petrels can theoretically be seen from any British and Irish shore, but there is a particular hotspot around Liverpool Bay, when autumn north-westerly gales can force them close inshore. On particularly blowy days there can be hundreds passing through the bay, giving birdwatchers spectacularly close views against a dramatic backdrop of high waves.

Worldwide there are an estimated 20 million individuals, distributed across various islands in the colder North Atlantic and Pacific Oceans. The largest colony, on Baccalieu Island off eastern Canada, holds more than 3 million pairs. There are four subspecies, although only the nominate form is widespread, the other three being found only on certain Mexican islands. Some Leach's Storm-petrels in North America have dark rather than white rumps.

This bird requires islands that are free of mammalian predators, and which offer soft soil in which burrows can be dug, or boulders, ruined buildings or rocky beaches with existing crevices that are suitable for nesting. Away from the breeding grounds the species is pelagic and often feeds well offshore. During migration it takes shelter close inshore if there are storms at sea, in bays and river mouths. In very severe gales it may be blown some distance inland. After major autumn storms a few Leach's Storm-petrels are often discovered over rivers or reservoirs, where they may remain for several days before attempting to reach the sea again. Less fortunate birds may be 'wrecked' inland far from water, where they stand little chance of survival. Wrecks can on occasion be very large – for example, 6,700 Leach's Storm-petrels were found inland following storms in November 1952.

BEHAVIOUR AND DIET

Leach's Storm-petrel, like other tube-noses, spends most of its time flying over the open ocean and covering great distances in its search for food. When flying slowly and looking for food, it 'patters' on the surface with its feet, but rarely lands on the water for any length of time. With proportionately longer and narrower wings than the European Storm-petrel, it is able to use dynamic gliding to reduce energy costs on longer flights. It is usually seen alone or in small groups.

Like the European Storm-petrel this species only comes ashore at night, because it is highly vulnerable to predators (primarily gulls, skuas and raptors) when on land. Leach's Storm-petrels that nest in the High Arctic breed in autumn rather than summer, in order to avoid having to come to land during perpetual daylight. On land they are slow and awkward, their legs being too weak to support more than a shuffle with much assistance from the wings.

The diet is composed primarily of zooplankton such as copepods and amphipods, as well as some small fish and shrimps. The birds may follow fishing boats and pods of hunting dolphins, and scavenge the floating remains of dead sea animals.

BREEDING

Leach's Storm-petrels return to their breeding grounds in late April or early May. They remain faithful to the same nest-site and mate year on year in most cases; they use calls to locate their partners, as they come ashore after dark. With an annual adult survival rate of 88 per cent, a few birds that have already bred will be without mates at the start of each breeding season, providing opportunities for new recruits of four or five years old to join the breeding population.

Newly formed pairs setting up home for the first time may need to dig out a new burrow – a laborious task. Digging is done with the feet, and sometimes birds dig a new side branch off an existing burrow, so two or more pairs may use the same entrance. The nesting chamber is sometimes lined with a few pieces of grass, brought in by the male.

The single egg needs about 40–42 days of incubation. The parents share the duties, alternating shifts of about three days. The bird that is not incubating goes out to sea to forage. A similar pattern continues for the first few days of the chick's life, but after that both adults leave it alone in its burrow during the day, returning in the night to feed it, by regurgitation, on a mixture of semi-digested fish and other prey remains, mixed in with stomach oil. The chick waits in its burrow, inactive and gaining weight rapidly, until by about 60 days old it is considerably heavier than an adult bird. The parents stop feeding it around this time, and it remains in its nest for a few more days before leaving in the night and making its first flight. It heads out to sea quickly to avoid predatory birds of the land and inshore sea.

A burrow or narrow crevice is needed for nesting, to keep the adults and chicks safe from danger.

MOVEMENTS AND MIGRATION

Leach's Storm-petrel is a full long-distance migrant, with British-breeding birds travelling south to spend winter over the tropical seas off West Africa. However, a few stay considerably further north. North American birds mainly overwinter off the coast of Brazil. There have been very few ringing recoveries of British-bred birds, making it difficult to do more than speculate on the migratory route. Hopefully satellite-tracking studies in years to come will add greatly to our understanding of this enigmatic bird's migratory habits.

A rare close view reveals this small petrel's distinctive steep forehead and neat nostril-tubes.

THE FUTURE

Conservation of Leach's Storm-petrel as a breeding bird in Britain comes down to the safeguarding of the birds on St Kilda. The island group today is the property of the National Trust for Scotland and is designated a World Heritage Site. It also holds a small military base. The only people who live permanently on the islands are defence personnel. Scientists, conservation volunteers and a few tourists visit in summer, but the islands are so far off the beaten track and so well protected that the risk to the petrels of serious disturbance by humans is very low. The islands have very few land mammals, and the arrival of rats, cats or other predators would have a disastrous impact on the Leach's Storm-petrels and other wildlife, so vigilance against this danger is constant. For example, when a fishing boat ran aground on Hirta in 2008, rat traps were set as a precaution. It is not known why the breeding population on Dùn has declined so steeply since Seabird 2000; possible causes are being studied.

One predator that cannot easily be kept away from St Kilda is the Great Skua, which has been increasing in number there. Although the petrels' nocturnal habits reduce the risk of predation, Great Skuas still kill thousands each year, and skua numbers have increased dramatically since the 1990s. There are similar issues at other Leach's colonies in northern Scotland. On a global scale, Britain's Great Skua population is of far more significance than its Leach's Storm-petrel population, so making any future decisions on whether and how to manage this situation will be complex.

STAYING ALIVE

Like European Storm-petrels, Leach's Storm-petrels are very long-lived for birds of their size, and reaching their thirties is not uncommon. Research into the species' cell biology has identified what may be the secret of their longevity. One known cause of ageing is the shortening of telomeres, lengths of 'spare' DNA at the ends of chromosomes that help to protect the functional DNA sections of the chromosome from degrading during cell division – one of the processes that causes ageing. In most species telomeres shorten with age, but in Leach's Storm-petrels (and, probably, other related species), they become longer with age. There are other factors at play, but it seems likely that this unusual physiological trait is key to the unusually long lifespans of the small petrels.

Other tubenoses

It is difficult to imagine a bird more out of place in Britain than an albatross. These huge, majestic seabirds breed in the Pacific and the Southern Ocean (surrounding Antarctica) – anyone taking a pelagic trip off New Zealand's South Island may see several species, dwarfing the many petrels and shearwaters in the scrum for chum. However, albatrosses are great travellers, and two species have reached British waters. Even more remarkably, one of these would surely have bred here, had it been able to find another **Black-browed Albatross *Thalassarche melanophris*** to mate with.

In 1967 the famous Gannet colony on Bass Rock was joined by an exotic visitor, a Black-browed Albatross, which sat incongruously among the Gannets and tried (without success) to court one of them. The bird was quickly (if unimaginatively) named 'Albert', and returned to the colony for the following two summers. Then, after a two-year absence, it appeared in a gannetry on Unst, Shetland, and made a home there for the next 23 years with just the occasional year off. The last year at Unst was 1995, but in 2005–2007 a Black-browed Albatross was in the gannetry on Sula Sgeir in the Western Isles, and many birdwatchers believe that this was the same bird. With wild albatrosses known to survive for more than 50 years, it is quite feasible, but a pity for the bird that it has not been able to breed.

There are nearly 20 other records of Black-browed Albatross on the British List, coming from as far afield as Kent, Devon, Yorkshire and Orkney, although these mostly relate to birds passing offshore. One of the smaller albatross species (sometimes known as mollymawks), it breeds on islands in the southern oceans and rarely strays above the Equator. Its world population is about 1.2 million, making it one of the most abundant albatross species, although it is still classified as Near Threatened because of serious declines in the late 21st century. Many birds are still accidentally killed each year by long-line and trawl fisheries.

On very rare occasions, albatrosses wander north to British seas, with the Black-browed Albatross our likeliest visitor.

The Black-browed Albatross looks superficially like a Great Black-backed Gull, with a white body and black wings and back. However, it has a very different shape, with preposterously long, narrow wings that make take-off very difficult but allow the bird to travel huge distances with minimal effort once airborne, via dynamic soaring. The bill is huge and orange, with a smooth outline (the nostril tubes are not prominent), and a dark area around the black eyes gives it a severe frowning expression.

The other albatross on the British List is the Atlantic **Yellow-nosed Albatross *Thalassarche chlororhynchos***. This bird is also a mollymawk, and looks quite similar to the Black-browed Albatross but has a dark bill with a narrow yellow stripe down the centre of the upper mandible. It breeds in the mid-Atlantic on islands including the Tristan da Cunha group, and is classified as Endangered. The sole British record was the first for the Western Palearctic and its circumstances were remarkable. A farmer found the bird, a subadult, waddling around on his driveway in Somerset. Although grounded, it seemed in good health, and was taken to a hilltop by the sea for release the following day. All seemed well as the bird took off and headed for the sea, but the following day it showed up again, this time much further inland on Carsington Water, in Derbyshire. It stayed for a short while before flying again, and was found later on during the same day on a lake in Lincolnshire. This bird's incredible journey across England happened at around the time of a spate of sightings elsewhere in Europe, including Norway, Sweden and the Faroe Islands.

The genus *Pterodroma* contains the 'gadfly petrels', medium-sized and relatively light-coloured tubenoses, most of which have a distinctive zigzag pattern across the uppersides of their very long, slender wings. Their taxonomy is still under debate but at present about 35 species are recognised, two of which have been recorded in Britain. **Fea's Petrel *Pterodroma feae*** is a Near Threatened but increasing species that breeds on the Cape Verde Islands and also the Desertas off Madeira (although these birds are sometimes split as a separate species, Desertas Petrel *P. deserta*). Its population is responding positively to mammal-eradication programmes on its breeding islands, and in most years one or two are seen in British or Irish waters, usually in the far south-west.

There are two British records of the **Black-capped Petrel *Pterodroma hasitata***, an Endangered, highly pelagic species that breeds in the Caribbean. This is one of the most distinctive *Pterodroma* species, with uniformly dark upperwings and back. However, Fea's Petrel is very similar to several other species that could theoretically occur in British and Irish waters, and because views are often distant, not every *Pterodroma* petrel seen can realistically be identified to species level.

With its blackish plumage, the Sooty Shearwater is distinctive, though could be confused with a dark-morph skua.

The Great Shearwater can be seen in huge numbers in the Bay of Biscay, but it is less often found in our waters.

Although the Manx Shearwater is our only breeding shearwater and the most likely species by far to be seen going by offshore, seawatchers occasionally record another shearwater species in greater numbers. The **Sooty Shearwater *Puffinus griseus*** breeds on various islands in the southern hemisphere, but is one of the world's great travellers, occurring at various times across almost the entirety of the oceans. Its migration follows a clockwise circle or figure of eight, to return to its islands at the start of the breeding season in November. Most of the migrants that skim our shores are southbound, seen between August and October. However, sometimes large numbers are seen heading north in the North Sea in autumn. These birds have been pushed off course on their southbound journey and are attempting to reach the Atlantic again to resume their preferred migratory route west of Britain.

Sooty Shearwaters are a little larger than Manx Shearwaters and have the same very long-winged, slim-billed outline, and shearing flight style. Their plumage is distinctive, being blackish-brown on the upperside and underside, with vague paler areas on the underwings. The only other seabirds as dark as this likely to be seen in British waters are dark-phase Arctic Skuas, which have broader, gull-like wings compared with the Sooty Shearwater's very long, narrow wings, while juvenile Gannets are much bigger birds with larger wings and a longer, stronger bill.

Satellite tracking has revealed that New Zealand's Sooty Shearwaters cover more than 60,000km per year in their epic migration, which takes them as far north as Alaska and across the whole Pacific region. This remarkable journey places them among the world's most prodigious animal travellers. The migrants make long stop-offs at certain areas, presumably those that offer the best feeding opportunities, but routes followed by individual birds vary widely.

The Sooty Shearwater breeds in huge colonies and the global population is very large, estimated to be some 20 million individuals in 2004. However, there are signs that it is undergoing a serious population decline, and it is classed as Near Threatened for this reason. It is threatened by introduced predators on some of its breeding islands, and by changes in fishery practices – especially the increase in long-line fishing. In some areas large numbers of chicks have been traditionally harvested for food for centuries, but this practice may prove unsustainable as overall numbers fall.

Another shearwater regularly seen in British and Irish seas is the **Great Shearwater *Puffinus gravis***. This bird has a dark upperside and white underside, like a Manx Shearwater, but also has a narrow paler collar and rump-patch, and its back and wings are a little lighter with pale feather edges creating a scaly appearance (although this is only obvious at close range). There is also a smudgy brownish patch on the belly. The darkest parts of the plumage are on the cap and tail. The species is considerably larger than the Manx Shearwater, but judging the size of a lone bird moving over the sea is very difficult.

With its light grey plumage and pale bill, Cory's Shearwater is one of the most distinctive of our rarer shearwaters.

Most sightings of Great Shearwaters are from headlands on the coasts of the Western Isles, Ireland and south-west England. Boat trips from the Isles of Scilly towards the Bay of Biscay can also be productive, and ferry crossings to northern Spain can produce numerous sightings. Like the Sooty Shearwater this bird breeds in the southern hemisphere in the austral summer, and migrates north to spend May through to August in the northern hemisphere. The migratory route follows a clockwise loop, with birds on their return southbound journey most likely to pass through European waters.

Great Shearwaters breed primarily on Tristan da Cunha (on Nightingale Island and Inaccessible Island) and on Gough Island. There are also thought to be a few pairs in the Falklands. The colonies are large and dense, with at least 5 million pairs on Tristan da Cunha and up to 3 million pairs on Gough Island. Many thousands of chicks and some adults are harvested each year by islanders, and while the population seems stable at present it is vital that monitoring continues to ensure that the harvesting level is sustainable.

Great Shearwaters are fast fliers, using dynamic gliding to cover long distances. They are not especially social, but due to their shared need to find feeding grounds, large numbers may be seen together, although this is very unlikely to happen from British or Irish coasts. They feed on fish and squid, as well as carrion from dead sea animals and offal scraps thrown overboard by fishing boats.

Cory's Shearwater ***Calonectris diomedea*** breeds much closer to the British Isles than either the Sooty or Great Shearwater, but is still an infrequent visitor to our waters. It breeds on islands in the Atlantic, with the main colonies being on the Azores, 1,360km west of Portugal. There is a separate population breeding in the Mediterranean. These birds are of the nominate subspecies *C. d. diomedea*, while the Atlantic birds are *C. d. borealis*. However, some authorities now split the Mediterranean birds as a separate species, Scopoli's Shearwater.

This large shearwater has grey-brown upperparts and white underparts, with a graduation between the two shades on its face and neck, giving it a less neat appearance than that of other shearwaters – it could be mistaken at first glance for a juvenile gull, but has the typical long, narrow shearwater wings and shearing flight style. There is a vague U-shaped white rump-patch, and the bill is noticeably yellow (rather than dark as in the other shearwaters recorded in Britain) and is rather thick. In October, after breeding, Cory's Shearwaters spread out across the North Atlantic to spend the winter foraging at sea, but they are not nearly such long-distance travellers as Great or Sooty Shearwaters and consequently only small numbers reach British waters. Most sightings are from headlands in south-west Ireland and south-west England. Any boat trip into the Bay of Biscay should produce many sightings, as the birds gather here to exploit prey brought near to the surface by upwellings from deep seas at the edge of the continental shelf.

The Mediterranean form of Cory's Shearwater, Scopoli's Shearwater, have occurred in British waters – sightings have been claimed on a few occasions. It is very similar to the nominate form, but has more extensive dark coloration on the underside of the primary feathers. This is, however, a very difficult field mark to spot on a fast-flying bird, and it is not really known how much individual variation there is between the two forms. The Mediterranean birds make up about 15 per cent of the total population of 270,000–290,000 pairs, but the species is declining. Illegal killing of adults and chicks is an issue (despite the species' protected status in the Azores and elsewhere), and some areas where the shearwaters breed have populations of introduced rats and cats. The RSPB is involved in habitat restoration and surveying projects in the Azores to improve the fortunes of Cory's Shearwaters and other seabirds there.

Very few of the bird species regularly occurring in the British Isles are considered threatened on a global scale. The most threatened species that does visit us is the **Balearic Shearwater *Puffinus mauretanicus***, a bird truly on the edge with a worldwide population of 9,000–13,000 individuals, and an IUCN status of Critically Endangered, meaning that extinction could be imminent without conservation action.

The Balearic Shearwater is a slightly larger than the Manx Shearwater, and is considerably browner, with a dusky breast and belly with no clear-cut division between that and the darker upperside. The undersides of the wings are whitish. The species is perhaps more likely to be confused with the Sooty Shearwater than with the Manx Shearwater.

British seas are important for the endangered Balearic Shearwater outside of the breeding season.

For land-based seawatchers, headlands in south-west England and Ireland offer the best chance of seeing Wilson's Storm-petrels.

The Balearic Shearwater breeds on the Balearic Islands and there are 450–900 pairs on each main island, adding up to about 3,000 pairs. However, at-sea surveys of non-breeding birds in 2011 and 2012 suggest that the total population may be substantially higher than would be expected from counts at breeding colonies, so there may be a large reservoir of non-breeding birds. Further studies are needed to clarify the picture, which may lead to the species' status being changed, although it is known that many breeding colonies have disappeared (for example, 60 per cent of those on Cabrera island have been lost since the late 21st century).

After breeding most of the birds leave the Mediterranean to spend winter in the Atlantic, primarily in the Bay of Biscay but also ranging further north. Around 800–1,200 are recorded each year on British and Irish shores, mainly in the south and west, but also in smaller numbers in the North Sea, up to the Scottish border. The causes of their decline are the usual factors affecting tubenoses – introduced mammalian predators on the breeding grounds, and their falling victim to long-line hooks and other fishing activities. It is also possible that overfishing in the Mediterranean is affecting breeding success. Like other shearwaters, the Balearic Shearwater only produces one egg per pair each year, so any recovery will inevitably be slow.

The **Macaronesian Shearwater *Puffinus baroli*** is a recently split species, formerly considered conspecific with the Little Shearwater *P. assimilis*. It is a very small shearwater, patterned like a Manx Shearwater with a pale face in which the dark eyes stand out, if views of it are clear. It breeds on Atlantic islands and winters in the southern hemisphere; there are a few British and Irish records from the mainland and boats.

One of the most sought-after species by keen seawatchers is **Wilson's Storm-petrel *Oceanites oceanicus***. There are about nine records a year in late summer to early autumn, although undoubtedly many more go by unnoticed. The challenge for birdwatchers lies not just in the species' rarity but in the difficulty of spotting such a small bird, flying very close to the waves and often obscured by them. Even from the best possible vantage point and at the perfect time of year, the birds will not be seen unless strong onshore winds push them close to land – normally they pass several kilometres out from the coast.

This tiny bird closely resembles the European Storm-petrel, but is a touch larger and has entirely dark wings. With very close views it is possible to see that the webbing between its toes is yellow, rather than black as in the European Storm-petrel. Like Sooty and Great Shearwaters, it is a southern-hemisphere breeder that migrates to northern seas during the austral winter. Birds seen from Britain and Ireland are on their return journeys, and are most likely to be seen from south-west Ireland and south-west England, and during boat trips that head south-west from those areas.

Globally, Wilson's Storm-petrel is an extremely abundant bird, with a population of an estimated 12–30 million individuals. It breeds on rocky islands and cliffs across the southern oceans, including the coast of Antarctica, where it is the smallest warm-blooded animal to breed. It is a typical storm-petrel in its feeding behaviour, moving slowly very low over the surface, and balancing on the breeze with raised wings and dangling feet. When flying more purposefully it flutters less and glides more than the European Storm-petrel. It feeds on mainly plankton but also targets floating fish remains.

A handful of other petrels have been seen from the British Isles, or found washed up dead on beaches. They are the medium-sized, mostly grey-and-white **White-faced Storm-petrel** *Pelagodroma marina*, the small **Madeiran Storm-petrel** *Oceanodroma castro*, which looks very similar to the European Storm-petrel but has a narrow rump-band rather than a square patch, and the larger, all-dark **Bulwer's Petrel** *Bulweria bulwerii*. **Swinhoe's Storm-petrel** *Oceanodroma monorhis*, which resembles a dark-rumped Leach's Storm-petrel, is another extremely rare visitor, but is notable for having been found to visit regularly in the breeding season and show signs of breeding behaviour. From 1989 to 1994 researchers were in Northumberland using tape lures to catch European Storm-petrels for ringing, and much to their surprise they caught three different Swinhoe's Storm-petrels in their nets. The species normally breeds in the north-west Pacific and winters in the Indian Ocean, so these records are quite remarkable, and illustrate the tremendous capacity of even the smallest seabirds to cover huge distances at sea.

A pretty and distinctive petrel, the White-faced Storm-petrel is an extremely rare wanderer to our seas.

Gannets, cormorants and relatives

The bird order Pelecaniformes traditionally included several families of large seabirds, including the gannets and boobies (known collectively as sulids), cormorants, darters, frigatebirds, tropicbirds and pelicans – all these groups share the trait of having webbing between all four toes. However, DNA studies show that things may be more complex than previously thought and the arrangement of the group is in the process of changing. In the British Isles there are two species of cormorant and one gannet, and a few related species have occurred as vagrants.

Gannets and boobies (family Sulidae) are graceful, long-winged and highly aerial birds, noted for their spectacular vertical plunge dives from several metres above the sea. This diving technique allows them to combine strong flying and relatively deep diving. Their breeding colonies can be very extensive and dense, with nests packed very closely together. Within this confusion of birds each pair is closely bonded, and when pairs reunite after time at sea they use calls and display to find each other and affirm their partnership. Worldwide there are ten species, of which one breeds in the British Isles and another has once been recorded in the English Channel.

The cormorants and shags (family Phalacrocoracidae) are slim and long-bodied birds with long necks and tails, and relatively long bills with hooked tips. Cormorants usually feed by surface diving, and use their feet for propulsion while under water. They have much less developed uropygial glands (for preen-oil production) than other seabirds, and their plumage structure also allows water to penetrate. They are therefore much less naturally buoyant than most diving species, allowing them to expend less effort when swimming under water, but requiring them to air dry their plumage after immersion by standing in a spread-winged stance. There are about 40 species worldwide, including some that are flightless, and others that use only freshwater habitats. Two species breed in the British Isles and another has occurred as a vagrant.

Frigatebirds (family Fregatidae) and tropicbirds (family Phaethontidae) comprise just five and three species respectively, all found in tropical seas. They are similar in that they are long-winged, very aerial birds and have elongated tail feathers when mature. Frigatebirds never swim on the sea, and feed mainly by stealing prey from other seabirds. Two species have been recorded in Britain. Tropicbirds feed in a rather tern-like manner, by hovering and plunge diving. There is one species on the British List.

A Gannet 'skypoints' to its mate, a display conducted prior to leaving the nest.

Gannet
Morus bassanus

On any coastline in Britain, if you spend a few moments scanning the sea through binoculars, there is a good chance of spotting a Gannet – or perhaps even several – flying low and strongly offshore. To really appreciate Britain's largest seabird, however, you need to visit one of its breeding colonies in summer, to watch hundreds of these magnificent birds coming and going from their nests on the flat tops of islands, stacks and rocky outcrops. To take a boat trip around one of the gannetries is an extraordinary experience – the foraging adult birds surround you, swirling overhead before plunge diving at high speed into the sea in pursuit of fish. Gannets are faring better than most British seabirds, with recent population increases. However, suitable breeding habitat for them is rather limited, so they are likely to always be rather localised and require rigorous protection.

INTRODUCTION

The Gannet is difficult to confuse with any other species, especially in the nearly entirely white adult plumage. The wingtips are solid black, but the rest of the wings are white, a pattern not found in any gull species. A close view reveals the delicate yellow-orange wash over the head and nape, and striking pale eyes set in blue bare skin. Positioned close to the bill-base, the eyes give the birds good binocular vision, allowing them to accurately judge distances during their dives and subsequent underwater chases. The body is cigar shaped, tapering at the ends with the long, pointed bill and tail, and the wings are long, quite narrow and pointed. The streamlined shape is an adaptation for plunge diving, as is the unusual bill anatomy with internal nostrils that open inside the bill, preventing water from being forced into the nostrils when diving. Juvenile Gannets have dark brown plumage, with a narrow pale rump-band and fine white speckles evident at close quarters. During the first four years of their lives they gradually acquire the white adult plumage over successive moults, with the head and underside becoming white ahead of the wings. Subadults look confusingly mottled, but the Gannet's size and distinctive shape makes it recognisable in all plumages.

Gannets use updrafts to hang on the air alongside cliffs, allowing them to survey their breeding colony at length.

With nearly 40,000 pairs, the gannetry on Grassholm Island in Wales is one of the most important in the world.

DISTRIBUTION, POPULATION AND HABITAT

There are only a few gannetries in Britain and Ireland, although some are very large. The only one in England, and one of very few mainland colonies, is at RSPB Bempton Cliffs in Yorkshire. In eastern Scotland there is a large mainland colony at RSPB Troup Head, and notable island colonies on the Bass Rock (from which the species gets the name *bassanus*), Hermaness and Noss in Shetland, on St Kilda (the largest colony, with more than 60,000 pairs), on Sula Sgeir in the Western Isles and at Ailsa Craig. In Wales there is a large colony on Grassholm Island off Pembrokeshire, and the largest gannetry off Ireland is on Little Skellig island, County Kerry. Some of these gannetries have more than 10,000 nests, and the grand total of nests in Britain and Ireland is more than 260,000.

Beyond the British Isles, Gannets breed on the Channel Islands, in northern Scandinavia, on Iceland, and across the Atlantic on Newfoundland and in adjacent Quebec. However, Britain and Ireland hold the bulk of the breeding birds (68 per cent). The total world population is 950,000–1.2 million individuals, a considerable proportion of which are birds of younger than breeding age. Non-breeding birds (adults in winter, and subadults year round) are found throughout the north Atlantic.

Gannets nest on flat or only gently sloped ledges and clifftops, on the mainland, on islands and on rock stacks. Suitable sites are well above sea level, inaccessible to mammals and have sufficient space for a good number of nests. A busy gannetry of many years' standing looks white even when no Gannets are present, because of the build-up of guano. Away from the breeding grounds Gannets are highly pelagic and wander well offshore.

After a successful dive, a Gannet gulps down its mackerel prey while simultaneously struggling back to the surface.

BEHAVIOUR AND DIET

Gannets are highly social, especially when breeding but also when foraging at sea. Even well away from breeding sites, it is more common to see them in small groups than alone. Where there is a good concentration of suitable fish prey, hundreds may gather. However, breeding colonies tend to have distinct foraging ranges that do not overlap with ranges used by neighbouring colonies.

With its narrow wings, the Gannet has high wing-loading and therefore sometimes has to work hard to get airborne and stay in the air. However, it is also expert at using wind currents to gain lift, and employs dynamic gliding when flying low at sea. The flight style thus combines powerful flapping with long glides when air conditions make gliding possible. When it is in feeding mode it tends to stay

HEADLONG PLUNGE

The ability to fly strongly and also dive deeply is a rare combination in seabirds – they generally tend to specialise at one and be mediocre or poor at the other. Taken to an extreme, these evolutionary pathways have resulted in the penguins, which dive deeper than any other birds but cannot fly, and the frigatebirds, which are superb aerial athletes but are unable to swim at all, let alone dive. The Gannet's special skill of diving steeply from a great height allows it to reach deeper water than is the case in other strong fliers like tubenoses, and various adaptations help to protect it from the physical challenges of such daredevil behaviour.

As mentioned above, the Gannet's nostrils are internal, so it breathes through its bill. This helps to prevent water entry, and the nostril openings inside the bill can also be closed off for a dive. The eyes are protected by sturdy nictitating membranes that close over the eyes during the dive. Finally, the bird's skull and neck are protected from the impact of diving by a system of air pockets under the skin, which have a cushioning effect rather like that of bubble wrap. Despite all this protection, diving is still a risky business for Gannets, particularly when diving en masse. Studies of the closely related Australasian and Cape Gannets have shown that collisions between two simultaneously diving birds are not uncommon and can result in serious injury and even death.

A Gannet halfway through its dive. It pulls its wings back just prior to impact.

higher, and circles and flaps as it scans the sea below before making its dive. When it begins its dive it tucks its wings in and tilts its head down to fall rapidly towards the water. Just before entry the wings are drawn right back behind it, shaping the bird into an attenuated and beautifully streamlined missile of a bird that punches deep into the water.

At the moment when a Gannet enters the water, the lenses in its eyes change shape from oval to spherical, to accommodate the different behaviour of light rays under the water. It swims under water with strong strokes of its large, webbed feet, although it may also use the movement of its half-folded wings to steer and corner. The dive may last 40 seconds or more, and can be as deep as 20m. Deeper dives tend to be made into a large shoal of fish that scatters in a panic, as the Gannet is not as fast or agile under water as a more specialised diving bird such as an auk. However, it also makes shallower, targeted dives to capture larger individual fish near the surface.

Food types favoured by Gannets include herring, mackerel, sandeels, sprats and other similar-sized fish. They rarely take non-fish prey, but do take fish remains thrown from boats. Each fish is swallowed as soon as it is caught, and on a single deep dive a Gannet may catch and swallow numerous small fish. After a dive it may rest on the surface before taking off.

Nesting pairs are evenly spaced, each pair defending a small territory.

BREEDING

As early as January Gannets return to their breeding colonies. Established breeding males normally return to the same nest year after year, and pair with the same female, which arrives slightly later. Young birds observe proceedings from the periphery of the colony and may build the beginnings of a nest there, but will not attempt to breed until they are four years old. They may try to hijack an existing nest if its rightful owner has not returned, but will not put up a fight if the true owner later comes back.

Gannet pairs display to each other in various ways to establish and reaffirm their bond. The two birds stand breast to breast with their necks stretched out and their bills pointing upwards; they turn their heads from side to side so that their bills gently rub together. Should a different bird encroach on a pair's territory, it will be vigorously rebuffed. The proximity of the nests to each other only seems to sharpen the birds' territorial instincts, and each pair fiercely defends its tiny patch, with the males fending off other males and the females doing the same with intruding females.

The first task in spring is to add material to the nest. Seaweed is the main component, but any other material the Gannets find may be used as well, including floating debris picked from the sea. Unfortunately, this often includes hazardous objects such as plastics, which can later fatally entangle a growing chick. The nest structure is held together with copious amounts of guano, but is still damaged frequently by the wind and the birds' activity, so nest maintenance is an ongoing chore. The largest nests are about 70cm across and 30cm high. The female lays her single egg sometime between late April and mid-June, with females nesting for the first time tending to lay later.

The nest is a heap of whatever material is easily collected, usually seaweed.

Incubation takes 44 days, and the task is shared between the sexes. Each shift lasts about a day and a half, the male's usually being a little longer than the female's. The average duration of shifts gradually shortens as incubation progresses – and does so dramatically just before hatching. Often, both birds are at the nest when the chick emerges, and one of them gives it its first meal of regurgitated fish soon after it struggles free of the egg.

Gannet chicks are rather unattractive, with baggy blue-black skin, closed eyes and very sparse white down. Over the first three weeks of a chick's life its down layer develops considerably, and by three weeks old it is entirely covered in thick, woolly-looking white down, and no longer needs to be brooded by its parents. Over the following eight or so weeks it grows quickly and gradually sheds its

In its first couple of weeks of life, the bare-skinned, reptilian-looking Gannet chick develops a coat of dense white down.

down, revealing its dark brown juvenile plumage. Throughout its long maturation its parents feed it between one and three times a day, and as it grows it begins to exercise its wings. When it is about 90 days old and weighs roughly a third more than an adult Gannet, its parents stop feeding it, and it spends another ten days or so on the nest, exercising its wings and gradually losing some of its excess weight, before leaving the cliff to fly down to the sea. Its ample fat stores sustain it for another two or three weeks before it begins to feed itself.

The weeks between parental abandonment and the development of fishing skills are a very risky time for young Gannets. Some starve, while others misjudge their first flight and blunder into other pairs' territories where they are violently attacked. However, if they overcome these hazards, Gannets can live well into their thirties.

HOT-FOOTING IT

Eggs need near-constant body warmth from a parent for the embryos to grow and develop. Most birds provide this by sitting on their eggs so these come into contact with their lower breast region, and they moult feathers from this area to create a bare brood patch, which transfers their body heat directly to the egg. An added bonus of this is that the shed feathers provide a soft lining for the nest.

Gannets do not develop a brood patch, perhaps because they cannot afford the loss in waterproofing that this would entail. Also, they already have an area of bare skin with a rich blood supply, in the form of their large, webbed feet. They place their feet on top of the egg when incubating, lowering their body on top to contain all the heat. The feet are also effective for losing excess body heat when necessary – both adult and nestling Gannets sit with their feet exposed to the air when they are overheating.

As it matures, the young Gannet's plumage changes from blackish to mostly white through successive moults.

MOVEMENTS AND MIGRATION

Gannets spend their winter in the North Atlantic. Many British and Irish Gannets do not wander far from their breeding grounds, but some travel a long way south, reaching waters off West Africa. Birds ringed in Britain and Ireland have been recovered from the entire West African coast as far as Senegal, and also on western Mediterranean coasts and around western Scandinavia.

THE FUTURE

This seabird has shown a marked population increase throughout the second half of the 21st century, and has established several new colonies. Numbers on Grassholm almost doubled in 1968–1998, while the recently established Troup Head colony grew from two pairs in 1985 to 2,787 in 2010. However, before this dramatic increase Gannets were heavily persecuted and exploited for centuries, so the change could simply be a recovery to 'normal' numbers. The most recent surveys reveal that the increase is continuing at a steady rate, with a 24 per cent increase in the UK overall in 2000–2012.

Gannets have a generalised diet and can forage long distances from their nests. This helps to protect them from the consequences of declining fish stocks in the North Sea, although of course their flexibility in this regard has limits. Their feeding habits make them less vulnerable to surface pollution than surface-diving species. However, toxins in the food chain can accumulate in their bodies and in some areas this has caused mortality.

Drowning on the end of a fishing hook is a fate that claims too many Gannet lives.

A hazard that will probably be noticed by anyone visiting a gannetry is that presented by plastic debris thrown into the sea – some adult Gannets are burdened by objects caught on their bodies, others die at sea from entanglement in discarded fishing nets, and a proportion of chicks die before fledging because their feet or wings are trapped in plastic material used in nest construction. Historically many chicks were harvested for food, and at some colonies, such as Sula Sgeir, the practice continues today. However, the 'take' is managed to ensure sustainability and is fully legal. Overall, the Gannet's position in Britain and Ireland at present looks very encouraging, and most colonies in other countries are also growing. As Britain and Ireland hold such a significant proportion of the world's breeding population, concentrated at relatively few sites, it is important that strong protection is maintained for those colonies.

Cormorant *Phalacrocorax carbo*

A long, sleek, dark shape powering over the wave crests on sturdy wings – the Cormorant is one of our most familiar seabirds, and is also a common land bird in some areas, since it lives and breeds inland as well as on the coast. It is a handsome if somewhat prehistoric-looking bird, with a sinuous grace and a habit of standing regally upright with its wings spread. However, it is not admired by all, and as a skilled fish catcher has come into considerable conflict with anglers.

INTRODUCTION

The adult Cormorant has uniformly blackish plumage with a blue gloss, except for the lower cheeks and upper neck, and a small patch on the flank, which are white. Mature adults develop a profusion of fine white filoplume feathers on the head and neck at the start of the breeding season, making them look 'grey-haired', while younger birds are browner and paler, especially on the belly. This bird has a torpedo-shaped body, tapering to a long tail; the neck is gracefully long, and the head is slightly peaked at the rear of the crown. The long, straight and sturdy bill has a pronounced hook at the tip for gripping slippery fish, and there is bare yellow skin around the bill-base that extends to surround the blue eyes. Cormorants look slightly goose-like in their strongly flapping flight, but are longer-tailed and shorter-necked than geese. They sit very low on the water, with their backs sometimes barely breaking the surface.

Cormorants are strong fliers, able to soar and glide thanks to their relatively broad and long wings.

Both at sea and inland, this is a gregarious species at all times of year.

DISTRIBUTION, POPULATION AND HABITAT

Cormorants can be seen around almost the entire British and Irish coastlines at any time of the year, and are also widespread in England inland. In winter their inland range extends to cover most lowland areas with suitable fishing waters in the rest of the UK and in Ireland. They are also present on most offshore islands. Colonies are not especially large, varying in size from between 10 and 400 nests. The Seabird 2000 survey looked at both coastal and inland sites and found 13,628 active nests in total in Britain and Ireland, 38 per cent of them in Ireland, 27 per cent in Scotland, 23 per cent in England, the Isle of Man and the Channel Islands, and the remainder in Wales. The coast/inland breakdown was 85 per cent on the coast (232 colonies) and 15 per cent inland (35 colonies).

The population increases in winter, when some immigrants visit from the Continent, to an estimated 41,000 individuals. Cormorants in Britain belong to two subspecies – the nominate, which occurs on Atlantic coasts, and *P. c. sinensis*, which is primarily found on the European mainland.

The Cormorant has an extensive global range, occurring on all continents except South America and Antarctica. Across most of its range it is known as the Great Cormorant, to distinguish it from other

WHICH SUBSPECIES?

The two subspecies of Cormorant that occur in Britain look very similar. They can sometimes be told apart by examining the shape of the gular pouch (the bare skin at the bill-base, on the throat). The angle of the pouch's rear edge relative to the bird's gape-line is smaller in *carbo* than in *sinensis*, although there is some mid-range overlap so not every bird can be identified to subspecies in this way. Those with an angle of 65 degrees or less are *carbo*, those with an angle of more than 76 degrees are *sinensis*. Even the most confiding parkland Cormorant may shy away from a birdwatcher wishing to press a protractor against its face, but it should be possible to measure the angle on a clear side-on photograph.

The subspecies question is important, because there is a common belief that the *sinensis* birds are recent immigrants to Britain, and that our inland-breeding populations are made up mainly or entirely of *sinensis*, while *carbo* sticks to the coasts. Those with anti-Cormorant interests have used these points as an argument that inland Cormorants are non-native and should be culled. The historical presence of *sinensis* is probably impossible to establish clearly, as we have only recently begun to understand how to separate the two forms, but there is some modern evidence that *sinensis* birds lead the way in colonising inland lakes and rivers. However, there are also many *carbo* birds at inland sites. DNA testing on Cormorants shot under licence at inland sites in England in 1997–1999 revealed that 70 per cent were of the *carbo* subspecies.

Because of its water-permeable plumage, the swimming Cormorant sits very low in the water.

resident cormorant species. It breeds in north-east North America and winters as far south as Florida. It is found in north-west and southern Africa, and patchily through central Asia and much of Australia. The world population is estimated to be 1.4–2.9 million individuals, and the overall population is currently increasing.

Coastal Cormorants breed on cliffs of all kinds, and also on islands in saltwater lagoons. Inland birds nest in trees that adjoin suitable fishing lakes, sometimes alongside Grey Herons. They feed in sheltered inshore waters and lakes and rivers of all kinds, even on town park lakes.

BEHAVIOUR AND DIET

Cormorants are often seen hunting alone, but they are fairly gregarious birds generally and can be seen travelling in small groups at any time of the year, as well as breeding in close-knit colonies. Large flocks fly in V formations like geese, an arrangement that allows the birds behind the leader to expend less energy, as they gain lift from the upwash of air produced by the leader's wingbeats. The birds spend much time resting near water, with wings held outspread if they have recently been immersed – their plumage is poorly waterproofed and has to dry off after each fishing trip to avoid risking hypothermia. There is also some evidence that the upright spread-wing posture aids the digestion, by allowing the sun's warmth to reach the bird's underside.

Waterlogged feathers need to be air-dried after a swim – the posture may also help speed up digestion.

Cormorants are ungainly on land, having short legs set close to the tail, but can perch on branches and walk in a shuffling waddle. They have fairly broad wings and can take off quite easily from land or water, and in the air can be surprisingly agile.

A Cormorant sits lower in the water than even the divers and grebes, and may just appear to be a serpentine head with no body as waves wash over its back. The head is usually held tilted slightly upwards. It makes a little jump out of the water when it dives, which may reveal the long tail (this is usually submerged when the bird is swimming). Under water, it swims with strokes of its broad, fully webbed feet. The dive lasts between 20 and 30 seconds, and the bird rarely goes below 10m – inland birds readily hunt in water much shallower than this. Sometimes several Cormorants work together to drive a shoal of fish to shallower water. This activity may attract aerial fish-eating birds such as gulls or terns.

Cormorants prey on all kinds of fish and often take relatively large ones, which they bring to the surface and may spend several minutes struggling to swallow. The benefits of eating a single, large prey item rather than fishing more frequently for smaller fare are clear for a bird that has to dry off after every fishing trip. At sea preferred prey includes flatfish, which the birds flush from the seabed, while inland Cormorants are often seen eating Eels and Pike, which are difficult to subdue but physically easier to swallow than equally large but shorter- and wider-bodied fish. Indigestible parts of the prey are later ejected in pellet form.

HOW MUCH FISH?

When you watch a Cormorant gulping down a huge fish that barely fits in its mouth and massively distends its neck once engulfed, it is easy to see why some people characterise this bird as 'greedy'. A look at various websites reveals some wild claims for the bird's eating powers. While most sensible people can easily dismiss some of these assertions – for example, it is clearly impossible for one 3kg Cormorant to put away 13kg of fish every day – it is still a common belief that Cormorants eat far more relative to their size than other fish-eating birds. In fact, the average intake is about 500g a day – relative to its body weight this is no more than is taken by a Grey Heron or Great Crested Grebe.

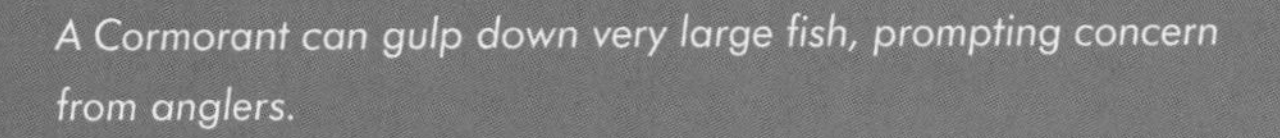

A Cormorant can gulp down very large fish, prompting concern from anglers.

BREEDING

Cormorants are nest builders, and this gives them more freedom in choice of nest-site than there is for seabirds that lay their clutches on bare ground. Some colonies are on cliffs, others on gravel islands in coastal lagoons, and inland they nest in large trees. The breeding season begins in around March, as the males establish territories and nest-sites, and attempt to attract a mate with a wing-waving display that reveals the white flank-patch. Sometimes birds pair with their previous year's partners, but many do not. Pairs do show various bonding behaviours, including a head-bobbing display and mutual preening sessions.

Nests may be reused if they are in a good state of repair, but birds that have lost their nests or males breeding for the first time (at three years old) must build new nests. The materials used depend on what is available locally, but nests are normally large heaps of vegetation, with plenty of sticks used in tree nests and seaweed in cliff nests. The nest has a soft lining of feathers and grass.

The clutch size is typically three or four eggs, but ranges from one to seven. The eggs are blue-green with a rather chalky texture, and the female begins to incubate them gradually when the clutch is partly complete, so there is usually an age difference within the brood. Both birds take turns to incubate the eggs, and they begin to hatch after 28–31 days. The hatchlings are fully altricial, with bare skin, closed eyes and very limited ability to move. They must be brooded by a parent by day

This species is an adaptable breeder, willing to nest on cliffs, in trees, and on shingle islands in reservoirs.

and night for about ten days, by which time the chicks have grown a coat of soft grey down and can sit up to beg for food. Their diet is regurgitated fish, which the parents place in the bill while the chicks are small, but as they grow the young birds take a more active role, inserting their heads into their parents' mouths to receive the food. At this point younger and weaker chicks may be unable to compete with their older siblings, especially in large broods, and some will starve.

The chicks fledge at 45–55 days old, and team up with other fledged youngsters. However, they continue to be fed by their parents for at least another month – the adults recognise their chicks' calls and are able to locate them within the congregation of youngsters. Both adults and youngsters leave the breeding areas when the chicks are about 90 days old.

MOVEMENTS AND MIGRATION

Cormorants are not great travellers, by and large. Inland breeders may continue to use areas near their colonies throughout the winter, while coastal birds tend to stay in inshore waters. Some of the greatest distances travelled, as revealed by ringing recoveries, include a bird ringed as a chick in Orkney and found 2½ years later in Norway; a chick ringed in County Wexford, which in the same year visited a boat in the Bay of Biscay 515km from its birthplace; and an Anglesey-ringed chick that at five months old had reached Italy, 1,541km away. By far the majority of foreign recoveries have been from France, but most ringed birds found in Britain were close to the sites where they were ringed.

Incubation is sporadic until the clutch is complete.

THE FUTURE

The Cormorant has experienced mixed fortunes historically. Persecution of coast-nesting birds was intense up to the mid-21st century, but new legislation protecting the birds allowed some recovery. Since the 1960s increased numbers of birds have also been breeding inland. In 1969–1988 the population increased by 9 per cent, and in 1988–2002 it rose by 10 per cent. However, counts in 2000–2011 indicated a decline of 11 per cent. This small overall change masks some very different patterns on a regional level – while some colonies (including those inland) are continuing to increase and new inland colonies are being established, northern coastal colonies, especially on the Scottish islands, have shown steep declines.

NON-LETHAL SOLUTIONS

There are many ways in which Cormorants' depredations on fisheries may be reduced without actually harming the birds. One of the most effective is the fish refuge, an underwater cage inaccessible to the Cormorants, in which fish can shelter. Most fish naturally prefer to avoid open water and stay near the margins of lakes and rivers, or among dense water vegetation, so thoughtful siting of fish refuges can greatly improve their chances of avoiding capture by hunting Cormorants. Other measures that have been tried with some success include deliberate disturbance, such as patrolling the water by boat, and using bird-scaring devices.

Because of inland birds' presence on lakes and rivers used by anglers, the Cormorant is one of the most controversial bird species in Britain. It can in some circumstances damage angling interests, and fishery owners in the UK may apply for a special licence to kill the birds, although only after non-lethal methods have been tried. The applicant must also show evidence that Cormorants are causing a serious problem. Up until 2004 the annual limit across all licence holders was 500 birds, but this has now been increased to 2,000, and the evidential requirements have been relaxed somewhat – the mere presence of Cormorants at a fishery is considered sufficient evidence that the birds are causing a problem.

Despite this recent change in the law, there remains much ill feeling towards Cormorants on the part of the angling community, and there is considerable pressure to add the species to the list of birds that can be killed under a general licence (with no restriction on numbers and no need to apply for licences). However, Defra's studies have found no evidence that large-scale control of Cormorants inland is needed to protect fish stocks.

Cormorants living on the coast and fishing at sea make up the vast majority of the British and Irish breeding population. While they do not arouse the same levels of ire as the inland birds, they face a range of threats, in particular changes in prey availability. They show a similar pattern of decline as other fish-eating seabirds, with some significant losses in colonies that use the North Sea. For example, the population on the Farne Islands fell from 238 breeding pairs in 1985–1988 to 144 recorded in the Seabird 2000 census (1998–2002), a decline of 39 per cent. Over the same time period the total Shetland population fell by 52 per cent, and the Orkney population dropped by 32 per cent, with some colonies disappearing altogether. Overfishing may explain these declines, although it is likely that some birds have shifted to different breeding grounds (including inland) rather than perishing.

Cormorants are also vulnerable to oil spillages, ingestion of plastic debris and poisoning from pollution. Entanglement in nets is another cause of mortality, as is deliberate persecution.

Both adults and chicks pant and puff up their gular pouches to lose body heat on hot days at the nest.

Flocks in flight often fly in V formation. Here a couple of pale-bellied young birds have joined the darker adults.

Shag
Phalacrocorax aristotelis

The Cormorant's smaller cousin, the Shag, is a much more exclusively marine bird than its relative, best searched for on cliffsides along northern and western coasts. It is a strikingly beautiful bird in its own way, with its serpentine grace and shining, almost oily-looking, green-glossed plumage. The two species can be difficult to tell apart, especially in subadult plumages. The Shag is rather similar to the Cormorant in its habits, and adopts the same stately cruciform pose to dry out its wings. It is present in many large mixed seabird colonies, although it does not nest in tightly packed groups and tends to stick to more sheltered, lower ledges, so is not as obvious as the throngs of Guillemots and Kittiwakes.

INTRODUCTION

The adult Shag has entirely black plumage with a green lustre. The fringes to the feathers are very slightly paler, giving the bird a subtle scaled look at close range. Young Shags are grey-brown, a little paler below especially on the chin. In the breeding season adults have a curly quiff of feathers at the front of the crown that can be raised or lowered. There is a small area of bare yellow skin around the gape-line, and this is brightest and most noticeable in summer. The eyes of adults are very bright green, while young birds have paler yellow-green eyes. The Shag's outline is similar to the Cormorant's, with a long tail, long, slim neck and long, hook-tipped bill. It is long bodied, with the feet set far back close to the tail. On land it either stands upright or rests on its belly.

A Shag in full summer plumage is panting to keep cool while incubating its clutch.

Shags look long-necked but rather heavy-bodied in flight, and tend to fly quite close to the sea surface.

DISTRIBUTION, POPULATION AND HABITAT

Shags breed along the coasts of Scotland, north-east England, Wales and south-west England, and around the entire coastline of Ireland, including on many islands and islets, wherever there is suitable habitat. The largest colonies, with more than 1,000 pairs, are on Foula in Shetland, on the Farne Islands, on the Isles of Scilly and on Lambay Island off County Dublin. In many areas, however, there are no obvious discrete colonies, and breeding pairs are spread out thinly along large stretches of coastline. The UK's total breeding population is some 27,000 pairs, with another 5,000 or so in the Republic of Ireland. This represents more than 40 per cent of the world's population.

The global distribution spans the Atlantic coastline from Finland to Morocco, and there are also populations along much of the Mediterranean coastline and parts of the Black Sea. The nominate subspecies is present in Europe, the subspecies *P. a. riggenbachi* in north-west Africa, and the subspecies *P. a. desmarestii* in the Mediterranean and Black Sea. Recent research suggests that birds in south-west Europe may also represent a separate subspecies.

Shags nest on rocky shorelines that offer safety from predatory mammals – on the mainland they generally use crannies or broad ledges on high cliffs, or boulder piles at cliff bases, but on islands they may use boulder beaches. They require enough ledge space to construct a quite substantial nest, and sometimes nest inside caves or on top of low rock outcrops adjacent to taller cliffs. Colonies

CORMORANT OR SHAG?

If you are in the south-east, or inland, it is highly likely that any cormorant-like bird you see will be a Cormorant rather than a Shag. However, a few Shags do wander further south and east, and occasionally even inland. The first thing to check is the general shape and impression of the bird. Cormorants look solid and robust, while Shags are markedly more slight in all respects. The Shag's bill in particular is much finer than the Cormorant's. In adult plumage Shags show no pale areas of plumage, while Cormorants have white areas on the face and flanks. Young birds are more difficult to tell apart, but young Cormorants have almost white bellies and chins, while young Shags have duskier undersides, and at all ages Shags show less bare skin on the face. The head shape is subtly different as well, with Shags showing a peaked forehead and Cormorants a peak at the back of the crown. Birds seen at a distance, however, may be impossible to identify for certain.

The very energetic take-off leap of a Shag as it dives helps it to propel itself under water.

are frequently in areas shared with other seabird species, and the Shags benefit from the increased vigilance of the other birds. They feed in seas over both rocky and sandy substrates, and are at ease in rougher water than are Cormorants, but still primarily fish in sheltered inshore areas. They are very rare inland but occasionally one is storm driven and takes refuge on a river or reservoir.

BEHAVIOUR AND DIET

The Shag has much in common with the Cormorant in terms of general behaviour. It spends most of its time resting on rocks near the water's edge, often with its wings spread to dry off from a recent dive. It is fairly social and non-breeding birds often rest up in small groups. When travelling at sea it usually flies close to the waves, and progresses with a steady, flapping flight, its wingbeats a little more rapid than the Cormorant's.

Shags on the water sit very low and hold their heads tilted upwards; when they dive they make a distinct forwards leap to help turn their bodies downwards. The water quickly penetrates their plumage and they reach neutral buoyancy at about 5m, compared with more than 30m for auk species. This significantly reduces energy costs while swimming under water. The Shag finds much of its prey at or near the seabed, and can dive to 45m. Typically dives are rather shallower, and last for 20–45 seconds, with pauses of 15 seconds between dives until a catch is made. The diet includes a wide range of fish types, including large numbers of sandeels in the breeding season. Individual catches tend to be of smaller fish than the Cormorant will take.

NAUGHTY NAMES

The Shag is perhaps the most unfortunately named British bird, and many birders have from time to time succumbed to the temptation to make a few off-colour Shag-related jokes when talking about their hobby. The origin of the bird's name, however, is quite innocent and relates to the bird's appearance rather than to any frisky behaviour. It comes from the Old English word *sceacga*, meaning a 'matted tangle' or 'tuft'. This describes the curious little crest that Shags develop in their breeding plumage, and is also the root of the usage of shag as a kind of rough carpet. The less salubrious meaning of the word 'shag' has a different origin, being related to the Germanic *skaken*, which simply means to 'move about'.

Shags use various materials for their nest, including land plants as well as seaweed.

BREEDING

As a rule Shags do not move far from their breeding grounds in winter, so there are birds present near their nests at all times of the year. In spring breeding activity begins in earnest, with males building new nests or renovating their old ones, and displaying to attract females by throwing their heads back and calling. The year-on-year fidelity between pairs is rather low, although it is higher in older birds, and some males even pair with two females. Shags can first breed from the age of two, although females are more likely to be older at their first breeding attempt than males.

When incubating, a Shag places its eggs on top of its foot webs, so the eggs are warmed from below as well as above.

The nest is made from plant material available locally, and therefore usually consists of seaweed and other floating vegetation and debris. No particular lining is added. The two birds share the work of nest building, but it is the male that selects the nest-site. The best sites are deep enough to accommodate a large nest, have straight-line access to the open sea while still being reasonably sheltered from wind and rain, and are high enough to be safe from big waves in the highest tides. At busy colonies younger, inexperienced Shags are more likely to build in suboptimal spots and are thus more likely to lose their nests to a big tide or severe weather.

More experienced and established pairs produce eggs sooner than newly formed pairs – overall the date of the laying of the first egg is any time between late April and late May. The completed clutch consists of usually three light brownish eggs, which the parents take turns to incubate, sitting on the nest and holding the eggs on top of their broad, webbed feet for maximum heat transfer. After 30 days the clutch hatches and one parent remains with the blind, helpless and naked chicks while the other goes to sea for food. The chicks' eyes open after a couple of days, and over the first two weeks of their lives they grow a coat of soft grey down, and their legs become strong enough to support them in an upright position. They can now thermoregulate well enough to be left alone, so both parents are free to forage for them.

Regurgitating food for the chicks looks a violent business, the youngster's head disappearing into the adult's throat. This parent is distinguishable by its uniquely numbered plastic leg ring.

The chicks can fly at 48 days, but their parents continue to feed them for some time afterwards, sometimes nearly as long as the period before fledging. There is some evidence that young Shags at times associate with a parent or parents for as long as 11 months. While the birds can begin to breed at the age of two or three and can live well into their twenties, many skip one or more breeding seasons over the years, at times when their body condition is not at its peak.

MOVEMENTS AND MIGRATION

Shags do not roam long distances at sea as a rule, and many spend winter very locally to their breeding grounds. Young birds are more likely to travel further than adults, as they explore the coastline for possible future breeding grounds. The greatest distance covered by a ringed and recovered Shag was by a bird ringed as a chick in Wales and found in Spain, 1,440km away. Other young birds hatched in the British Isles have reached Denmark and Norway, and there have been many recoveries from the north coast of France. A number of French-ringed young birds have also been found on British coasts.

Shags are susceptible to being 'wrecked' by heavy weather at sea in the autumn, and may find their way to inland water bodies. This usually involves singletons, but on occasion groups of nine or more have appeared together. Sometimes they settle very well in their new surroundings and make a lengthy stay. Others are less fortunate and perish away from water.

In juvenile plumage, the Shag is a dusky-brown bird, with a darker belly than same-aged Cormorants.

THE FUTURE

After dramatic increases up until the 1960s and in some areas the 1980s, Shags are in a period of decline, with a 26 per cent loss in 2000–2011. Numbers were falling before this as well, with the Seabird 2000 survey indicating that significant losses have occurred in Scotland, especially in the north, and in Ireland, since 1988. There have been small overall increases in England and Wales over the same timescale. The Seabird 2000 total of 32,306 nests was slightly down on that recorded in the Operation Seafarer surveys of 1969–1970 (33,876 nests), but dramatically lower than the results of the Seabird Colony Register census in 1985–1988 (42,970 nests). It should be noted that as a result of the Shag's habit of nesting in tucked-away crannies and caves, a certain proportion of nests is bound to be missed when taking a census.

Some ornithologists believe that the increase in Shag numbers in the last century was partly in response to a general cessation of harvesting the birds for food. This practice was widespread in some areas. Published in the 1950s and 1960s, Lilian Beckwith's books about her life on the Isle of Soay, near Skye, made several references to the islanders catching and consuming Shags. Another cause of the increases was the reduction in persecution that followed the introduction of increasingly strict legislation protecting the birds and their nests from destruction.

It is difficult to assess whether the Shag population is now falling back to a more natural level following an unusually productive spell in the late 21st century, the result of natural fluctuations, or whether the decline is cause for concern. There have been a number of distinct events over recent years that have impacted on Shag numbers in the British Isles. A prolonged stormy spell in late winter 1994 killed an estimated 3,000–5,000 adult Shags in north-east England/south-east Scotland. The *Braer* oil spill off Shetland in 1993 also took a heavy toll on their numbers. With 857 dead Shags picked up from beaches after the disaster, they outnumbered the next most frequent victim (Black Guillemot) four to one. The annual seabird count on the Farne Islands has revealed some marked changes in the last few years, with 1,059 pairs of Shags in 2007 but just 492 in 2008, and similar counts in the following three years. However, in 2012 things picked up considerably, with a count of 965 pairs (some other species on the islands showed similar increases in 2011–2012).

This bird's population trends are unclear – careful monitoring will be needed to determine future conservation action.

Shags remain vulnerable to oil spills and other forms of pollution. Climate change may affect them negatively, by changing sea temperatures and therefore changing the abundance of fish populations in the Shags' foraging grounds, and also by increasing the incidence of storms that affect the survival of nests in spring. Some adult Shags drown in discarded nets, while for the eggs and chicks predation by introduced rats is an issue in some areas – Shag numbers increased sharply on Ailsa Craig following rat eradication there – and overfishing is another potential cause of decline. However, Shag numbers on the Isle of May have remained fairly stable despite the existence of a large fishery there in the 1990s, extracting more than 100,000 tonnes of Lesser Sandeels in some years. To fully understand why our internationally important populations of this species are in decline, more study is necessary, to investigate hatching and chick survival rates, the likelihood of young birds returning to their natal colony to breed, and also adult survival rates and breeding productivity at the various colonies.

Other related species

There is only one British record for the **Double-crested Cormorant *Phalacrocorax auritus***, a North American species that looks very similar to our own Cormorant. The bird was found on a pond in Billingham, Cleveland, in January 1989, and stayed until April that year. It was fortunate that the bird remained so long, as establishing its identity was a real challenge, requiring detailed observation and much specialist knowledge. The Double-crested Cormorant has a similar ecology to our Cormorant, being a bird of both coastlines and inland waters.

No sulids besides the Gannet have been admitted to the British List, but in 2007 a **Masked Booby *Sula dacylatra*** was reported in the English Channel, 35km south of Portland Bill. This species breeds on tropical islands worldwide except for the eastern Caribbean, and has been seen in Spanish waters on three occasions, so future sightings in Britain may occur in years to come. The Masked Booby is a little smaller than the Gannet, with more black on the wings, a black tail and dark skin around the eyes.

Two species of frigatebird have been recorded in Britain. These large, highly aerial seabirds rarely range north of the Equator so their occurrence here is quite remarkable. The first was an **Ascension Frigatebird *Fregata aquila***, found on Tiree in the Hebrides in 1953. The unfortunate subadult female bird was pulled ashore in a fisherman's landing net – it was exhausted and in poor condition, and

The Double-crested Cormorant lacks its 'crests' when not breeding and is very similar to our Cormorant.

Magnificent Frigatebirds are only likely to reach British seas if caught up in extreme weather conditions.

died overnight. A second Ascension Frigatebird was photographed on Islay in July 2013, only to disappear before the twitchers arrived. It was thought for many years to be a **Magnificent Frigatebird** ***Fregata magnificens***, but was reidentified in 2003. However, it would only be two more years before a genuine Magnificent Frigatebird arrived in Britain. A farmer found the bird, an adult male, in a field in Shropshire. It, too, was exhausted and in poor condition, although it survived for a month under veterinary care at Chester Zoo before succumbing. Another Magnificent Frigatebird was found on the Isle of Man in 1999, again in a very poor physical state, and it too died in captivity, this time after ten months of care.

The tropicbirds are no longer considered to be closely related to the other species in the Suliformes, and are placed in their own new order, Phaethontiformes. One species, the **Red-billed Tropicbird** ***Phaethon aethereus,*** was added to the British List after a bird was seen off the Isles of Scilly in 2001. The bird was seen and photographed by a yacht crew, and has previously been found as a tideline corpse, in Suffolk in 1993. More recently, the remains of a **White-tailed Tropicbird** ***Phaethon lepturus*** were found on a beach in Cumbria in January 2013. Both birds were thousands of kilometres north of their usual range.

A beautiful and distinctive bird, the Red-billed Tropicbird has only once been observed alive in British waters.

Phalaropes

The bird order Charadriiformes is large and diverse, comprising several seabird families including the skuas, gulls, terns and auks. Another key family within the order is Scolopacidae, which contains about 90 species of wader. These are mostly long-legged and long-billed birds, which feed by picking from damp ground or probing into it. Many of them are associated (in winter at least) with coastal habitats such as saltmarshes, estuary mudflats and even beaches, and they can and do swim from time to time. However, those in only one group, the phalaropes, are specialist swimming and water-feeding birds.

There are three species of phalarope in the world, the Red-necked, Grey and Wilson's Phalarope *Phalaropus lobatus*, *P. fulicarius* and *P. tricolor*. The first two of these breed around the Arctic Circle, while Wilson's Phalarope breeds only in North America. All three species migrate long distances southwards after the breeding season, to spend the winter around tropical oceans. They feed primarily on small insects and, at sea, on zooplankton and tiny crustaceans, which they pick from on or near the water's surface, or upend to capture. Wilson's Phalarope sometimes feeds on land, but the other two species rarely do so.

Phalaropes are small and rather hyperactive waders with slender, straight bills. In breeding plumage they are colourful, with red, brown and black-and-white patterns, and the females are larger and brighter than the males. This switch of the usual arrangement in sexually dimorphic birds is reflected in a role reversal in the breeding season. The female defends a territory to attract a male, then guards her mate from other females until she is ready to lay her eggs. Once the clutch is complete the female abandons the male. He incubates the eggs and raises the chicks alone, while the female either repeats the process with a second mate, or begins her southwards migration early. Juvenile phalaropes and those in winter plumage are greyer and drabber than breeding adults, with no red tones.

All three of the phalaropes are on the British and Irish Lists. The Red-necked Phalarope is a very rare breeding bird on lochs in the far north of Scotland. The Grey Phalarope occurs regularly offshore as a passage migrant. Wilson's Phalarope is a very rare vagrant, most often seen in autumn following gales from the west.

A unique bird, the Grey Phalarope is the only wader that truly qualifies as a seabird in Britain.

Grey Phalarope
Phalaropus fulcarius

Both the Grey Phalarope and the Red-necked Phalarope are true seabirds when not breeding, travelling and feeding far out to sea. Although the Red-necked Phalarope breeds in Britain and the Grey does not, only the Grey Phalarope is regularly seen offshore around Britain and Ireland. Most of the birds that pass our shorelines are never seen by any birdwatcher, but a few are driven close enough inshore to be recorded by seawatchers, and in extreme weather one or two may be forced inland.

Known as the Red Phalarope in America, this bird is only likely to be seen in Britain in grey, non-breeding plumage.

The world population of this species is 1.1–2 million individuals, and the nearest breeding birds to the British Isles are on Iceland. There are also populations on Svalbard, as well as along the Siberian coast, and in the far north of North America. The species is known as the Red Phalarope in America because of its breeding plumage, which is mostly deep brick-red. This plumage is unlikely ever to be seen in Britain. The birds we see are mostly juveniles transitioning into first-winter plumage, and they are white below with a mixture of brown and pearly-grey on the upperside. By late October most are

in full grey-and-white winter plumage, and the only dark areas they show are the flight feathers and a bandit mask over the eyes.

Between 200 and 400 Grey Phalaropes are recorded in Britain each year, the majority occurring along the east and south coasts of England, between September and November, when the birds are migrating south. The distribution pattern suggests that most birds recorded here originate from further east, but there are not enough ringing records to ascertain this and no tracking studies have taken place to date.

Only in windy weather are Grey Phalaropes likely to be pushed close enough to the shore for good views as they go by. In the aftermath of easterly gales, a few storm-driven birds may take refuge in sheltered bays or on coastal lakes, and rest and feed before resuming their journeys. They can then be observed at leisure and, with luck, at very close quarters as they seem completely unafraid of people. The feeding style they employ is distinctive. They are strong swimmers, having broad lobes on each of the three forwards-facing toes that join at the bases to form webs. They often spin in rapid circles to stir up the water and bring tiny aquatic creatures to the surface, where they can be snapped up.

Phalaropes have a distinctive swimming stance, very buoyant and upright, with a spinning action when feeding.

Skuas

Birdwatchers are always delighted to see a skua, but to other seabirds skuas are a much less welcome sight. These gull-like birds are pirates and predators, specialising in stealing food from less agile birds, and as well as eating fish and carrion they sometimes prey on eggs, chicks and even adult seabirds. Where penguins are the focus of wildlife documentaries and even children's animated films, skuas are invariably cast as the sinister villains of the piece, but they are in fact charming, charismatic and very beautiful birds in their own right.

The skua family Stercorariidae is a small group with seven species. All are now classified in the same genus, *Stercorarius*, but there are two distinct groups. The large, mottled brown, short-tailed skuas, mainly restricted to the southern hemisphere but represented in the north by the Great Skua, were formerly placed in the genus *Catharacta*. The exact taxonomy of the various very similar types of large brown skua that live in the southern hemisphere is still hotly debated – some authorities recognise three species, others four or five. However, in the north things are more straightforward, with four clearly distinct species – the Great Skua and three smaller species, the original *Stercorarius* skuas. These three species all have elongated central tail feathers in adult plumage, and are known as jaegers in North America. All four northern skuas have been recorded in the British Isles, two of them as breeding birds.

Skuas nest in colonies on open ground, and are less protected from predators than cliff-nesting seabirds. They compensate for this vulnerability by having a coordinated and very robust defensive response whenever danger threatens – anyone wandering too close to nesting skuas should be prepared for an aerial assault and would be advised to walk the long way round.

Outside the breeding season skuas migrate to warmer water and may be seen flying past headlands in autumn, and again in spring as they return. The two rarer species are keenly sought after by seawatchers, especially in spring, and may even be seen chasing and harrying other seabirds while migrating. Skuas may also make overland crossings on their migration, and when they make stop-offs they may rest and loaf on beaches, causing consternation among the local gulls and shorebirds, before moving on.

A dark Arctic Skua overflies its moorland breeding habitat.

Arctic Skua
Stercorarius parasiticus

With the languid grace of a bird of prey, the Arctic Skua creates a distinctive silhouette as it cruises along a shoreline in search of smaller seabirds to menace. Like other skuas it is a kleptoparasite, specialising in stealing prey from other birds, although it also hunts for itself from time to time. In a British context it only breeds in Scotland, but it is a long-distance migrant and may be seen offshore anywhere around the British Isles during the spring and autumn. Unlike the Great Skua, the Arctic Skua is in serious trouble as a British breeding bird, having experienced steeper declines than any other assessed seabird species since the year 2000.

INTRODUCTION

A little larger than a Common Gull, this bird gives an impression somewhere between a gull and a falcon, with its long, pointed wings and agile flight. When perched it can look deep chested, long winged and strangely small headed, and has a gull-like bill with a hooked tip. Adult Arctic Skuas have elongated central tail feathers, extending as needle-like spikes beyond the otherwise fairly short and square-cut tail. There are distinct colour forms of adult birds. Pale morphs (uncommon among the breeding population in Britain) are dark grey-brown above and dusky off-white below, with a neat dark cap, a smoky flush on the flanks and breast-sides (sometimes forming a complete breast-band), and a golden flush around the head and neck. Dark morphs are entirely dark grey-brown, but both forms show a white wing-flash formed by the bases of the primary feathers. Many birds are intermediate between the dark and light forms. Juveniles also occur in different colour forms, ranging from very dark to quite light, warm brownish-buff with strong dark speckling. They have short tails and less noticeable wing-flashes than adults. Subadults show developing central tail-feather spikes and plumage intermediate between adult and juvenile.

This skua has a falcon-like outline with a distinctive tail-shape.

Low-growing heather or grassland offers good all-round visibility, so breeding skuas can spot any predators long before they get near the nest.

DISTRIBUTION, POPULATION AND HABITAT

The Arctic Skua's breeding stronghold in Britain is on Shetland, where just over half of all breeding pairs are found. Orkney holds the bulk of the remainder, and there are much smaller colonies on the Western Isles, some of the inner Hebrides and northern mainland Scotland.

This is a very widespread species, breeding on northern coasts across the whole of Eurasia and North America (where it is known as the Parasitic Jaeger). In Europe it has strong populations in Scandinavia and Iceland, but not anywhere further south. Outside the breeding season it migrates to the southern hemisphere, and because of the wide sweep of its breeding range and its willingness to migrate overland, it has been recorded in well over 100 countries worldwide.

The British population during the Seabird 2000 census was 2,136 breeding pairs. No more up-to-date full census data is available at the time of writing, but some key colonies have suffered steep declines – for example, a 2010 survey on Orkney found 380 nests compared with 720 during Seabird 2000, and Shetland counts in 2012 of colonies on Foula, Fair Isle, Hermaness, Mousa and Noss found 65 nests altogether, compared with 218 in 2000. If such declines are typical, the total British breeding population may now be fewer than 1,000 pairs. This represents a small fraction of the world population, which is estimated at 500,000–10 million individuals (see below).

Arctic Skuas breed on damp moorland and high rough grassland with pools, mainly near the coast but sometimes several kilometres inland, especially on the Scottish mainland. Elsewhere in the world this skua also breeds on mossy tundra. Away from the breeding grounds it may be seen along any coastline.

MAKING IT COUNT

This bird (as with the other northern skuas) is one of the more difficult seabirds to count, for a variety of reasons. Its preferred breeding habitat is rugged and difficult to traverse, often on boggy ground. The nests within a colony are usually quite well spaced, and some pairs nest away from any others. Nests are often situated in hollows, and when one bird is on the nest and its mate is absent, the incubating bird sits low and tight, so may easily be overlooked. Knowledge gleaned from earlier surveys helped the researchers for Seabird 2000 to refine their search methods and produce the most accurate count to date. However, applying thorough and effective search techniques across the species' entire breeding range, much of which is very far from human habitation, would be a mammoth undertaking. This explains the very open-ended current estimate of the world population, which is based more on the extent of suitable habitat than on any actual counts. However, going by counts in Scotland, there is a chance that this species could be in real trouble on a global scale, so more detailed population assessment in other countries is needed.

Ganging up on an Arctic Tern means these Arctic Skuas stand a great chance of stealing its prey, though only one of them will win the booty.

BEHAVIOUR AND DIET

This bird is most often seen in flight, and has a flight style that recalls a large falcon such as a Peregrine, with strong, fast wingbeats interspersed with wheeling glides. When in pursuit mode it exhibits tremendous agility, matching its victim turn for turn. On the water it floats buoyantly, looking quite gull-like, and can take off quickly. On land it walks and runs with ease. It is usually seen alone when not at its nest, although migrants may go by in small parties. On the breeding grounds it is fiercely aggressive to all intruders.

The Arctic Skua targets smaller seabirds than itself when seeking a free fish meal, and harries them relentlessly. Favourite targets are Kittiwakes and terns, but it also attacks auks. Some Arctic Skuas may feed exclusively on meals stolen from other birds, applying the same kleptoparasitic methods to whatever species they encounter in their southern hemisphere wintering grounds. Others take a more diverse diet, feeding on fish scraps discarded by boats, washed-up carrion and live prey, mainly in the form of small fish like sandeels that they catch by splash diving. They may also kill small seabirds on occasion, although they are less likely to do so than are Great Skuas or indeed the larger gulls. Among tundra-breeding populations, small mammals such as lemmings can form an important component of the diet, as do eggs and chicks taken from other birds' nests.

BREEDING

Adults return to their breeding grounds in May, and seek to nest at the same site with the same mate as in previous years. However, if a bird is late to return it may find that its mate has found a new partner in the interim, as young birds yet to breed (three years old or older) will be present on the colony fringes, looking out for opportunities like this. As is the case with other species exploiting the short northern summer, time is of the essence when breeding. Often a significant number of pairs do not attempt to breed.

The nest-site is a rough scrape on the ground, often somewhat protected by hummocks, but equally often allowing a view of the sea in colonies on the coast. In it the female lays a clutch of two

JOIN THE CLUB

Established breeding pairs of skuas are much more likely to be successful than new pairs, so they tend to stick together and jealously guard their nest-sites. It is therefore difficult for young birds to be recruited into the breeding population. Skuas are long lived, and if an established bird loses or leaves its partner there will be no shortage of potential new mates from among the youngsters. Young birds can of course pair with each other, but then they face the problem of finding themselves a nest-site, as the best spots will already be occupied. Young Arctic Skuas therefore need to be close to the breeding birds to stand the best chance of securing the ultimate prize – an older bird, already experienced, with a well-placed nest-site but no partner. At large colonies especially, the young birds establish 'clubs' on the edge of the main colony, where they can watch over proceedings. Some also set up and defend territories, and even create 'practice' nest scrapes. This rehearsal of breeding behaviour helps to improve their chances of one day successfully doing it for real.

eggs, which are olive or greyish with dark mottling. The pair alternates incubation shifts, and the off-duty bird remains nearby when not feeding, ready to launch attacks against intruders if necessary. At a large colony the coordinated response of several pairs of Arctic Skuas to a potential predator is a dramatic sight. Although smaller than Great Skuas they are even fiercer, and because of their greater agility they are more likely to score a 'hit' on their target, although the force of an attack is less severe just because the birds are smaller. However, some individual skuas prefer to sit tight on their nests when danger threatens, rather than risk leaving the eggs.

Incubation takes 25–28 days and begins when the first egg is laid, resulting in an age difference between the chicks. They hatch fully covered in down and with open eyes, and by three days old they will have left the immediate nest scrape to find separate hiding places within their parents' territory. They remain well hidden until a food delivery arrives from either parent, whereupon they approach to beg for food. In years when food supplies are low, the younger chick may not survive.

The young skuas can fly at about 25–30 days old, but need to be fed by their parents for another two to five weeks. They are vulnerable to predators in those first few weeks after their first flight, as they are still weak and badly coordinated on the wing, and cannot easily escape more powerful predatory birds. Attacks by Great Skuas on newly fledged Arctic Skuas can have a significant impact on breeding success.

At a couple of weeks old, a young Arctic Skua is losing its baby down and wandering away from the immediate vicinity of the nest.

An Arctic Skua in Shetland. Here the species is struggling, with recent poor breeding success.

With no fish catch to protect, an Arctic Tern turns the tables on an Arctic Skua, chasing the larger bird away.

MOVEMENTS AND MIGRATION

Arctic Skuas leave their breeding grounds in July or August and begin to migrate southwards. The peak month for sightings on British and Irish coasts is September, when this is usually the most numerous skua seen on seawatches. Many of the birds that go past our shores are breeders from further north and east, where pale morphs are more common, so on migration a higher proportion of pale birds are seen than on the breeding grounds in Scotland. Often Arctic Skuas are seen in the midst of a strong migratory passage of feeding terns, seemingly appearing from nowhere to dash in among the terns and try to chase one down.

The majority of birds in Britain track down the Atlantic coasts of France, Spain and Africa, to spend winter off the southern African coasts. There are rather few ringing recoveries but most show this pattern – however, some birds cross the Atlantic to eastern North America or eastern Brazil. There have also been a handful of British-ringed recoveries from within the Mediterranean, including the far eastern edge, from western Greenland and Scandinavia. Most remarkable of all have been single recoveries from the Democratic Republic of Congo and Sudan, in Africa, both thousands of kilometres from the sea.

COLOUR CHOICE

There has been much research into the mate choices of Arctic Skuas, in relation to plumage colour. Because both pale and dark morphs exist (in varying proportions) across all populations, there must be some survival and/or breeding advantages for both colour forms, otherwise one would have disappeared or be greatly outnumbered by the other. There is evidence that females prefer to mate with darker males, and males with paler females, which would help maintain the balance. The variation may be due to the fact that in some breeding habitat types pale birds are more conspicuous but in others the reverse is true.

Great Skuas cause multiple problems for breeding Arctic Skuas, so the latter are quick to mob and chase their larger cousins.

THE FUTURE

Like the Great Skua, the Arctic Skua showed a population increase from the 1960s to the 1980s. The Operation Seafarer survey in 1969–1970 found 1,039 nests, compared with 3,388 in the Seabird Colony Register in 1985–1988. A reduction in persecution may be a factor behind this increase, but improved coverage and better counting methods might be of more significance. As these factors were improved again for Seabird 2000, it is possible that the decline in Arctic Skuas between the Seabird Colony Register and Seabird 2000 was even greater than the 37 per cent recorded.

The decline to 2,136 nests between the Seabird Colony Register and Seabird 2000, and the even steeper decline that now appears to be underway, are considerable causes for concern. It is very likely that Arctic Skuas are feeling the impact of reduced sandeel stocks. Sandeels may not be a directly important food source for the skuas, but it is for the terns, Kittiwakes and other birds that they parasitise. In 2004, Arctic Terns on Shetland experienced near-total breeding failure, and as a consequence so did Arctic Skuas. When food supplies become harder to find, the chicks are at risk not only from starvation, but also from predation as the adults are forced to spend more time away from the nesting area.

Deliberate persecution was historically quite significant – Arctic Skuas were believed to block access for livestock to the best grazing grass, and were also regarded as a nuisance because of their strongly territorial behaviour. However, both legal protection and a softening of attitudes generally has helped to reduce persecution to very low levels. Loss of breeding habitat due to agricultural changes has also been of little significance in recent years.

Conflict with Great Skuas, on the other hand, is a likely contributory cause of the decline. Great Skuas have increased nearly as rapidly as the Arctics have declined, and use similar habitats. Where the two species breed side by side, Great Skuas can usurp Arctic Terns from their nest-sites and sometimes even kill the adults, thus removing experienced breeding birds from the population. For a species with such low breeding productivity, especially among newly formed pairs, the loss of several established adults in a season can have a serious impact on a colony. Great Skuas also kill fledgling Arctic Skuas and this can account for considerable numbers. The adult Arctic Terns are unable to effectively defend their offspring once the young birds are flying.

Measures to help Arctic Skuas are tied in to helping the other seabird species on which they depend. Any action that helps to improve breeding success for terns, auks and Kittiwakes in northern Scotland should have a knock-on effect for the skuas. Establishing Marine Special Protection Areas in places where these birds forage for sandeels will be a key part of a recovery plan. Studies are ongoing to explore the interactions between Great and Arctic Skuas. Because it is thought that Great Skuas prey more on other seabirds when fish supplies are scarce, improving the availability of fish for all seabirds is likely to reduce predation of Great Skuas on Arctics.

Great Skua
Stercorarius skua

This imposing, powerful bird – often known among birders by its Shetland name, Bonxie – breeds in the far north of Scotland and north-west Ireland, and unlike most of our breeding seabirds is expanding its range. It is an important species from a conservation perspective because it is an uncommon European endemic, and Scotland holds more than half of its entire global breeding population. Seeing Great Skuas away from the breeding areas is an unpredictable business, but they always add extra excitement to a seawatch when they come powering past a headland, perhaps in pursuit of a hapless gull or Gannet.

INTRODUCTION

At first glance the Great Skua looks like a large, dark juvenile gull. Although only the size of a Herring Gull it gives the impression of a larger bird, because of its very stocky, thick-necked, broad-bodied build and broad wings. The plumage of both adults and juveniles is primarily dark brown with darker mottling and fine paler streaks. There is often the suggestion of a dark mask around the eyes. Both the upperside and underside of the wings show a broad and obvious white 'flash', formed by white bases to the primary feathers. The bird has a strong, hook-tipped dark bill, and dark legs and feet. It flies strongly with powerful downstrokes, and while it may look quite ponderous in straight-line flight it can be fast and very agile when chasing other birds.

The prominent white 'windows' in a Great Skua's wings readily distinguish it from similar-sized young gulls.

The largest of our skuas is a formidable bird with a very powerful and muscular build.

DISTRIBUTION, POPULATION AND HABITAT

Great Skuas are thought to have first colonised Scotland in around 1750. For many years thereafter they remained rare and were almost entirely restricted to Shetland. This island group still holds more breeding birds than any other area, but there are also now strong populations on Orkney, and growing numbers in the Western Isles and on the north-west mainland. The first breeding pairs in Ireland are thought to have arrived in the 1990s, and there is now a small established population on islands off the west coast.

The Seabird 2000 survey counted 9,634 pairs in Scotland, a 26 per cent increase since the mid-1980s, and by now there are probably in excess of 10,000 pairs. The world population is estimated to be 16,000 breeding pairs and, including non-breeding birds, 48,000 individuals. This makes the Great Skua one of the least abundant seabird species to breed in Britain.

Beyond the British Isles Great Skuas also breed in Iceland, Norway, Svalbard and the Faroe Islands. They all move south in winter, mostly no further than the Atlantic coasts of France and Spain. Their breeding habitat is rocky islands, or open grassland or moorland near the coast, and outside the breeding season they live at sea but may also use beaches to rest. Occasionally they are storm-driven inland and head for a water body such as a reservoir to rest before returning to the coast.

LOCAL CONCERNS, GLOBAL ISSUES

The recent increase in our Great Skua population has raised some concerns, because Great Skuas are predators on other seabird species, as well as taking other fare such as fish and carrion. On St Kilda Great Skua numbers have increased from ten pairs in the 1960s to 240 pairs in 2000, and these Great Skuas prey almost entirely on other birds. As their numbers increase, they may be placing considerable pressure on populations of other species, in particular Leach's Petrel, which has its British stronghold on St Kilda. Remarkably, a few skuas have learned to hunt the petrels at night, thus circumventing the petrels' main defence against avian predators.

The situation on St Kilda raises some awkward conservation questions for the future. The islands' populations of Leach's Petrels are in steep decline and Great Skuas may be at least partly responsible. Great Skuas are generally on the increase and are spreading southwards, while in a British context, Leach's Petrels are extremely rare away from St Kilda. However, worldwide there are some 20 million Leach's Petrels and only 48,000 Great Skuas, so considered on a global scale the skuas are more 'important' than the petrels. Working out how best to conserve and protect both species is a complex juggling act, although at present it is not clear whether the skuas are a key cause of the petrels' decline.

Any small animals are fair game to this highly predatory bird, and carrion is also readily taken.

BEHAVIOUR AND DIET

Great Skuas are not particularly gregarious, and although they breed in colonies they are territorial and will not tolerate other Great Skuas near their nests. When foraging they are usually seen alone, but several may assemble at a rich food source – a washed-up seal carcass, for example. They are strong fliers and are comfortable walking and swimming as well.

This species' piratical behaviour is well known and can be seen at seabird colonies. The skua targets a bird on its way back to the nest having caught fish at sea, and chases it relentlessly, trying to exhaust it and drive it down to the sea. The skua will actually grab hold of its victim's wings, tail or bill in the air, or bomb it as it tries to rest on the sea's surface. The only way the victim can make its pursuer leave it alone is by dropping its catch, or disgorging the contents of its crop in the case of species that swallow prey at sea. Even Gannets, which are larger and twice as heavy as Great Skuas, regularly fall victim – the skua is much more manoeuvrable and can easily make it impossible for a Gannet to get back to its nest. Smaller birds such as auks and Kittiwakes are at high risk of being killed by the skua if they do not give up their catch quickly, and sometimes even if they do.

Conflict between Great Skuas over breeding grounds is commonplace, especially in areas where the population is growing fast.

Great Skuas can pick prey or morsels of floating carrion from the water's surface. They also catch live fish by splash diving, and may catch small diving seabirds in the same way. In some areas discards from fisheries make up an important part of their diet.

The diet of the Great Skua is diverse and shows some marked regional variation, with St Kilda birds feeding mainly on other birds, while Shetland birds eat a much higher proportion of fish. The birds readily eat carrion and fish offal thrown from boats, and take any unattended eggs and seabird chicks that they find. While breeding birds tend to have a distinct specialisation in one prey type or another, those on migration are more opportunistic.

BREEDING

Great Skuas arrive on their breeding grounds in April, and established breeders nearly always pair with the same mate again. This saves valuable time, which is important for these northerly breeders as they must make the most of a rather short annual window of ideal breeding conditions, and established pairs are more likely to breed successfully than newly formed pairs. One study found an annual 'divorce' rate of 6.4 per cent – a partnership is three times more likely to end when one of the pair dies than by divorce. When a pair does divorce, the separation is more likely to be initiated by the female

On the nest an incubating skua is inconspicuous, but it remains alert and ready to chase off would-be predators.

than by the male, and females sometimes invade a pair's territory and drive out the resident female. From the age of four or five, young Great Skuas may attempt to breed, but most first-time breeders are at least seven years old.

The birds arrive on territory with ample fat stores that have been accumulated on their wintering grounds, and weigh about 25 per cent more than they do after their eggs hatch. This extra weight helps to tide them over through the early weeks of the breeding season when food supplies are rather low. Each pair quickly establishes a small territory, which is defended against other Great Skuas. However, the colony acts as one when danger threatens. Walkers who stray near nesting Great Skuas are dive-bombed, and may be kicked at, pecked or even knocked down by the birds as they dive, with the attacks intensifying closer to the nests. Non-human predators and grazing mammals (which could trample nests) are treated in the same way, and this is a highly effective deterrent. Great Skuas sometimes breed alongside Arctic Skuas and the two species can make an effective team against predators. However, the benefits are limited for the smaller species, as Great Skuas may also displace Arctic Skuas from their breeding habitat, and prey on their eggs and chicks. This is a serious problem for Arctic Skuas as they are declining dramatically across Scotland.

It is wise to be wary if crossing a moorland where these birds are nesting, as they do not hold back on discouraging intruders.

In the nest, a scraped-out hollow on the ground, the female lays two eggs. These are grey-greenish marked with darker spots. Incubation starts when the first egg is laid, so it will hatch ahead of the other, and that chick will have a distinctly better chance of survival than its younger sibling, especially in lean years. The female carries out the bulk of the incubation, while the male guards the territory and forages for food that he will share with his mate. The eggs take 26–32 days to hatch, with colder weather extending the incubation period. The newly hatched chicks are semi-precocial, having a full coat of down and open eyes but limited mobility. The female guards them at this time while the male commutes to the sea to find prey, returning regularly to regurgitate food for the youngsters and the female. After a few days the two chicks are strong enough to leave the nest scrape and tend to go their separate ways within the territory. When a parent returns with a crop full of food the chicks rush over, begging with shrill calls and pecking at the adult's breast to stimulate regurgitation.

By the age of 44 days the young skuas are strong on their feet, have spent much time exercising their developing wings and are ready to make their first flight. They make exploratory journeys over the next three weeks or so while still being fed by their parents, but after that the young birds fend for themselves and both they and the adults leave the colony. Young skuas live at sea for several years.

Chicks are semi-precocial, being alert and downy but with limited mobility for their first few days.

MOVEMENTS AND MIGRATION

By August most Great Skuas have already left their breeding grounds and are on their way south. September and October are the months when they are most often seen by seawatchers, as they move along the coastline and sometimes fly close inshore when there are strong winds at sea. By December sightings in British waters have become rather rare.

There have been many recoveries of British-ringed birds, and these show that most migrants spend their winter off the Atlantic coasts of France or Spain. However, a sizeable proportion continue down the west coast of Africa, a few reaching as far as Cote d'Ivoire, and others enter the Mediterranean, as far as Italy. Some cross the Atlantic, and reach the western shores of South America. There have also been recoveries in Iceland, Greenland, Newfoundland and north-east Canada, across north-west Europe (including some well inland) and as far as north-west Russia. Many of these wanderers are young pre-breeding birds. The oldest ringed bird to be recovered was 36 years old, and there have been several other recoveries of birds in their thirties.

Opposite. It is not unusual to see pairs cruising together alongside seabird cliffs, looking for unattended chicks.

THE HYBRID THEORY

Great Skuas have been known to hybridise with Pomarine Skuas in the wild. This seems surprising, as the two species are quite dissimilar in appearance, the Pomarine Skua being distinctly smaller and much more like the Arctic Skua than the Great Skua in its plumage. However, comparisons between the mitochrondrial DNA of Great and Pomarine Skuas are remarkably similar. Not only is the Pomarine Skua more closely related to the Great Skua than to the Arctic Skua, but the difference between them is the smallest ever measured between two vertebrate animal species. Current reasoning suggests that the Pomarine Skua may be a rare example of a 'hybrid species' – the consequence of extensive hybridisation between Great and Arctic Skuas, resulting in a large hybrid population that went on to settle into a stable new species. It was the discovery of the close relationship between Pomarine and Great Skuas that led to the reassignment of all the large skuas into the genus *Stercorarius*.

In the past, Great Skuas were heavily persecuted because of conflicts with shepherds, although the birds pose no risk to healthy sheep.

THE FUTURE

This species' success as a breeding bird in Britain seems to be largely the result of an increase in discards from commercial fisheries – an ideal and convenient food source for them. Breeding Great Skuas have been persecuted in the past, because they were believed to present a danger to sheep-farming interests – territorial birds may attack sheep to keep them from trampling the nesting areas, and thus block access for the sheep to areas of grassland. The birds started to gain legal protection in 1900 and are now fully protected, but some illegal shooting does still occur. Population crashes on Fair Isle as recently as 1987, 1993 and 1998 have been partly attributed to shooting, as shot birds were found locally. Although this colony has continued to grow, it has not done so at the same rate as other colonies established at about the same time, suggesting that the birds have not recovered strongly from the string of declines.

The proportion of the diet that is made of other seabirds has increased generally in Scotland. There is also considerable year-on-year variation, with more birds taken in years when fish are harder to find – which also means that less food is available to the skuas in the form of discards at fisheries. So in years when other seabirds suffer reduced breeding success because of shortages of important fish (especially sandeels), their problems may be compounded by increased Great Skua predation.

At present Great Skua numbers are still increasing, but it seems likely that limitations in available food, whether fish or seabirds, will eventually curtail the population growth. A few of the older and larger colonies are already showing signs of this. Other potential causes of decline include pollution – as a long-lived, top-level predator this species accumulates large amounts of pollutants in its body – but the impact of this on factors like survival and breeding success has yet to be established.

With such a small world population, and a large proportion of it based in Scotland, the Great Skua is of high conservation priority. Protecting it on the breeding grounds and preventing illegal killing is important, but its impact on other seabird species, in particular Leach's Petrel and the Arctic Skua, must also be investigated to provide a full understanding of the dynamics before it can be decided whether any measures should be taken to curb the Great Skua's excesses.

Other skuas

The lovely **Long-tailed Skua *Stercorarius longicaudus*** is a rare visitor to British shores. This is the world's smallest and most elegant skua species, with a slight build, very graceful flight and, in adult plumage, greatly elongated central tail feathers that flutter as it flies. Adults exist only in a pale morph, which is light grey-brown above and paler below, with a black cap and yellowish cheeks, and in flight they show only a hint of a white wing-flash formed by the shafts of the outermost couple of primary feathers. Juveniles range from very dark to very pale, but are always more grey toned than juvenile Arctic Skuas, with slightly longer central tail feathers than the rest of the tail. Identification is still far from easy, especially as skuas are often seen going by some distance offshore.

This skua breeds on tundra in the Arctic regions of Eurasia and North America (where it is called the Long-tailed Jaeger), and has a global population of 150,000–5 million birds. The closest breeding populations to our shores are in Norway. These birds migrate to the southern hemisphere in winter, and pass British and Irish shores on their outbound and return migratory journeys. However, they are very comfortable travelling far from land and sightings from the coast are infrequent. Offshore gales may drive them towards our shores, and watchpoints along the east coast and the Hebrides are the best places to look for them.

Smallest and most elegant of the skuas, the Long-tailed Skua in adult plumage is unmistakable if seen well.

Unlike other smaller skuas, the Long-tailed Skua only has one colour morph (pale) when adult, although juveniles are more varied.

The Long-tailed Skua has never been known to breed successfully in Britain. On rare occasions, however, one may join a Scottish breeding colony of Arctic Skuas or set up home on other suitable breeding habitat and spend the summer here. In 1980 a pair of Long-tailed Skuas was at a site in Angus and Dundee, where it held territory and produced a clutch of eggs – but sadly the nest was predated. More recent signs of breeding behaviour include a single bird on Shetland in 2010, which held territory for nearly two months. The Long-tailed Skua could be a potential future colonist of Britain – its distribution pattern in Europe is similar to that of various other bird species which breed in northern Scotland in very small numbers. Its chances of success here are hard to assess, but its ecology is significantly different from that of the declining Arctic Skua, as it feeds primarily on small mammals and is in fact the least parasitic of the skua species.

The largest of the 'small skuas' is the **Pomarine Skua** ***Stercorarius pomarinus,*** which is known to be a very close genetic cousin to the Great Skua despite looking more similar to the Arctic Skua (see page 120 for more details on the relationship between Great and Pomarine Skuas). It is larger and sturdier than the Arctic Skua, and in adult plumage shows elongated central tail feathers that are thick rather than needle-like, and look twisted, with bulbous tips. The unusual shape of these feathers has led to their being nicknamed 'spoons'.

This is a powerful and highly parasitic skua, which chases other birds to steal their prey and sometimes actually kills its victim if it manages to force it down onto the sea. It also feeds on fish that it scavenges from the water's surface or shoreline, or catches for itself in a shallow splash dive. It has a steady, strong, gull-like flight, and a buoyant gull-like swimming style. Light-morph young birds are very gull-like in appearance as well, and could be confused with juvenile Herring Gulls at first glance. However, a close-up look reveals their more long-winged, long-tailed, shorter-billed and small-headed proportions.

Adult Pomarine Skuas may be of dark or light morphs, with light morphs predominating among sightings from British shores. Both forms show a white wing-flash that is less obvious than the Arctic Skua but more prominent than the Long-tailed Skua. Juveniles are of pale, intermediate or dark morphs, and all forms are rather sooty-grey toned.

The Pomarine Skua breeds on Arctic tundra in the far north of Eurasia, and round across North America (where it is known as the Pomarine Jaeger). The closest breeding birds to Britain are in Russia. The birds migrate south to winter in tropical seas, and on their journey may pass British and

Irish coasts. In autumn the east coast seems the best area for sightings, but in late spring headlands along the south coast of England, the Solway Firth and the Western Isles are real 'hotspots' for seeing adults returning towards their breeding grounds, often going by in small flocks. At these sites day counts of passing Pomarine Skuas can reach triple figures if the wind conditions are right.

Several large skuas breed in the southern hemisphere, all of them very similar to the Great Skua in appearance. They are the **Chilean Skua *Stercorarius chilensis***, the **South Polar Skua *S. maccormicki*** and the **Brown Skua *Stercorarius antarcticus***, which is sometimes split into three separate species. None of these birds would naturally occur anywhere near British waters. However, there are a few records of birds that do appear to belong to these species.

In 2001 a large, solid dark brown skua was found injured on St Agnes in the Isles of Scilly. The bird was taken into care to be treated and rehabilitated, and was eventually released into the wild. During its spell in captivity the identity of the 'Scilly Skua' was hotly debated. Some considered it to be a dark Great Skua, while others thought it was more similar to a Brown or South Polar Skua. Feather samples were obtained for DNA analysis, and these strongly suggested that the bird was a Brown Skua, with an outside chance that it was a hybrid between Brown and South Polar Skua. The two species are known to hybridise in the wild.

The following year another similar-looking skua was picked up injured in Glamorgan. This bird was also treated and feather samples were taken. Once again, the bird's DNA indicated that it was a Brown Skua, with a chance of it being a Brown x South Polar hybrid, but its measurements were a good match for the Falkland Islands form of Brown Skua *S. a. antarctica*, which is sometimes split as a distinct species (Falkland Skua).

Numerous specimens of Brown and South Polar Skuas have since been DNA tested and the results of the tests have been compared to those of the two 'mystery skuas'. Both the Scilly and Glamorgan skuas are now thought most likely to belong to the latter species. However, because identification to species level cannot be absolutely ascertained, neither the South Polar nor the Brown Skua has been added to the British List. It is clear, however, that these birds were certainly 'southern skuas' of some kind or another and had made truly remarkable journeys to end up where they did.

The Pomarine Skua is a beefy bird, approaching Great Skua in size and bulk.

Gulls

For many of us, gulls or 'seagulls' are the archetypal seabirds. They are a constant presence in seaside towns, loitering on the beaches, nesting on the rooftops, and eagerly dealing with dropped chips and ice creams on the pavement. They are versatile generalists in their feeding behaviour and habitat use, adept at quickly exploiting advantageous conditions, and several species have a strong presence inland, as well as (or instead of) on the coast.

It therefore comes as a surprise to many to hear that most of our breeding gull species are in decline, some of them severely. They have been affected, as have other seabirds, by overfishing and by reduced availability of scraps and offal from fisheries. In the case of Herring and Lesser Black-backed Gulls, breeding inland is a relatively new development, but the inland colonies seem to be faring better than those on the coast. It is testament to these birds' adaptability and intelligence that they have been able to successfully make the transition to a very different way of life. There are also some gull species that remain highly specialised, such as the very pelagic Kittiwake which rarely goes further inland than its sheer nesting cliffs, and Sabine's Gull which is almost a pole-to-pole migrant, only likely to be seen by British birders when it passes our coasts offshore in autumn.

Gulls all belong to the family Laridae, which is part of the large and diverse order Charadriiformes. They are closely related to both skuas and terns. Besides the seven gull species that regularly breed here, many others occur as regular or occasional visitors. Some of these come to us from North America, some from the High Arctic and some from mainland Europe. They present birdwatchers with some of the biggest identification problems of all, particularly the subadult birds in their various intermediate plumages. Some birders find the challenge irresistible and spend their spare time visiting refuse tips to catalogue the visiting gulls. However, this 'larophilia' is not shared by all, and Herring and Lesser Black-backed Gull are listed on the general licence, as a result of which anyone can humanely destroy them or (in the case of Herring Gull) their nests if they are found to be posing a threat to public health.

Worldwide there are about 55 species of gull, with representatives breeding on all continents. The majority belong to the genus *Larus*, with the rest split over up to ten other genera depending on taxonomy. About half of the world's species are on the British List, with a few others considered to be potential future vagrants to Britain or Ireland.

Black-headed Gulls breed inland as well as on coasts, favouring marshy wetlands.

Kittiwake
Rissa tridactyla

The Kittiwake is a part of every classic cliff-face seabird colony in Britain. It is also the chief noise-maker in the seabird symphony, with a raucous three-note call that gives it its name. The most marine of all our gulls, it very rarely wanders inland and shows a number of adaptations to a more ocean-going life than that of its relatives. Although still numerous as a breeding bird here, it has declined rather sharply in recent years and in the UK is Amber listed as a species of conservation concern. It is one of only two species in the genus *Rissa*, and is sometimes called the Black-legged Kittiwake to distinguish it from its rare Pacific relative, the Red-legged Kittiwake.

INTRODUCTION

This is a particularly beautiful and elegant gull, and the only British-breeding species to show no brown in its plumage as a juvenile and subadult. It is a little larger than the Black-headed Gull and in breeding plumage has pure white underparts and head, a pearly-grey back and wings, and neat jet-black wingtips with no white spots. Its legs are black (occasionally reddish) and rather short for a gull. The claws on the three forwards-facing toes are strongly curved to help with grip, but the hind toe is reduced or absent. This, along with the short legs, gives it a less horizontal stance than the postures of other gulls when perched. The bill is yellow and the eyes are dark.

Juvenile Kittiwakes have the same colour scheme as adults, but with additional black markings – a spot behind the eye, a rather narrow collar, the tail-tip and a zigzag line from tip to base on the upperside of each wing. Birds in this distinctive plumage are sometimes known as tarrocks. Once they moult in their first summer and acquire second-winter plumage they are more adult-like but still have much extra black in the wing, along with a soft grey 'boa' around the back of the neck in place of the juvenile's black collar. Winter adults have a similar 'boa' and also a dark eye-spot. In flight the birds look relaxed and graceful, and show a slightly notched tail shape. The grey in the primary feathers has a translucent look in good light, contrasting with the inner parts of the wing.

Against the light, Kittiwakes show markedly translucent primary feathers.

Ledges on buildings are just as good as cliff ledges, as long as the building is very close to the sea.

DISTRIBUTION, POPULATION AND HABITAT

The Kittiwake has one of the most extensive ranges of a cliff-nesting seabird, with colonies on suitable cliffs on all coastlines. The largest populations are on the east coasts of Scotland and northern England, on Orkney and Shetland, and in north-west Scotland. The largest gaps in its distribution are along south-west Scotland and north-west England; from Lincolnshire to South Norfolk; and along the south coast of England, where there are just a few scattered colonies. The most recent full count (Seabird 2000) found 415,995 nests across Britain and Ireland, 33,000 of these in Ireland. However, there is evidence of a substantial decline – nearly 50 per cent – between 2000 and 2012.

Worldwide the Kittiwake is very numerous, with a total population of 17–18 million individuals. It breeds on northern coasts and islands across Europe, Russia and North America, with a few colonies on the Atlantic coasts of France and Spain. In winter it lives at sea, mainly in the north Atlantic and north Pacific near its breeding colonies, and only rarely further south.

A TALE OF THREE (AND A HALF) TOES

The Kittiwakes on British and Irish coasts, and on other North Atlantic coasts, are of the nominate subspecies. There is a second subspecies, *pollicaris*, which breeds on the North Pacific coast. The most noticeable difference between the two subspecies is a curious one – *pollicaris* has a 'normal' gull foot structure with three long, forwards-pointing toes, joined by webs, and a smaller unwebbed hind toe with a claw, but in *tridactyla* the hind toe is absent or reduced to a tiny bump with (usually) no claw. The subspecies names reflect this difference – *tridactyla* means 'three-toed' while *pollicaris* comes from the Latin word *pollex*, meaning 'thumb' (although strictly speaking it should have had a name derived from *hallux*, the word for 'big toe').

Why one subspecies should require a functional hind toe while the other does not is a mystery. It could be that for birds nesting on the sheerest cliffs, which have more need to cling than to walk, the hind toe is not needed and would even be an impediment. There is regional variation in hind-toe size across both subspecies, which would support the idea that hind-toe size is linked to differences in nesting habitat type.

Any kind of cliff type is used by Kittiwakes, although they specialise in very sheer faces with narrow ledges, which are inaccessible to mammalian predators. They also use artificial structures if they are on the coast and have suitable ledges; such structures include piers, bridges and gas platforms. When feeding they forage both inshore and well out to sea. They may join other gull flocks to loaf on beaches but are very rare inland, and their occurrence there is usually as a result of being storm driven.

BEHAVIOUR AND DIET

During the breeding season Kittiwakes are highly gregarious, and even when not directly engaged in breeding activity they rest together in groups on nearby breakwaters or on the beach. At other times of the year they may be seen alone or in small parties. They are more aerial than most gulls and also spend much time sleeping on the sea's surface.

This gull takes mostly live prey, which it catches near the sea's surface. The most common feeding method involves making shallow plunge dives from a metre or two above the water. The gulls may also catch prey by picking it from the surface in flight or while swimming, although calm water is needed for this strategy to be successful. Swimming birds may make rapid turns, like those of a phalarope (see page 107), to stir up the water and bring tiny swimming organisms to the surface. Kittiwakes sometimes follow fishing boats to take scraps thrown overboard, where their extra speed and agility helps to give them the edge over large gulls, Gannets and Fulmars – although they may then have their prize stolen from them by one of the more powerful species.

The most important food items for Kittiwakes, especially in the breeding season, are small sandeels. They also feed on other small shoaling fish such as sprats, Capelin and immature herring. When not breeding they take a higher proportion of scavenged food, including fishery discards and washed-up carrion. Small prey picked from on or near the surface may include shrimp and other small crustaceans. Kittiwakes have been noted feeding near sewage and power-station outflows, to which they are attracted by the large numbers of fish that gather there.

Like most gulls, Kittiwakes are social and are often observed in good-sized flocks.

THE WONDER OF WOO

Mating successfully while balancing on a narrow ledge is a challenge for would-be Kittiwake parents.

Courtship behaviour in birds serves the main purpose of allowing both sexes to gain information about each other. This information includes the immediate 'state of mind' of a bird – how receptive it is to forming a pair bond, and how suitable a partner it would make. Ritualised behaviours allow birds to demonstrate their general fitness and condition, and also their ability to fulfil the requirements of successful breeding, which include tasks such as guarding the nest and providing food for chicks.

Displays noted in Kittiwakes include the 'greeting ceremony', which begins when one bird returns to the nest. The birds give the loud 'kitt-i-waaake' call to each other both during the 'fly-in' and when they are together on the nest. Then they stand close together and continue to call, bobbing their heads up and down in unison with their bills open and touching. This behaviour appears to signal recognition and receptiveness. In bonded pairs the male often goes on to feed the female, regurgitating food into his throat which the female takes by inserting her bill into his open mouth. By feeding his mate the male can both demonstrate his provisioning skills, and help his partner to gain fat stores that will prepare her body for forming and laying eggs.

BREEDING

Throughout February, March and early April, Kittiwakes return to their cliffside breeding colonies. Males take possession of their previous year's nest-sites and call to advertise them and themselves to females. Often, previously successful pairs re-form, while young birds look for a nest-site and/or partner from the age of three, but may not successfully pair up for another two years or more. Males whose mates have died over winter are more likely to attract a new partner if their nest-site is nearer the centre of the colony. Established pairs nesting at the centre of the colony begin their clutches earlier and are more likely to be successful than newer pairs and those on the colony edges. Pairs spend much time together at their nest-sites, developing their bonds with various courtship behaviours.

Kittiwakes build nests with a mud base, and the main body of the nest is usually made with seaweed. They collect this from the shoreline, find it floating on the sea and pick it from structures that project into the sea. Existing nests are likely to require some renovation after the winter months, so much of the colony's activity in the last few weeks before eggs are laid in May is focused on gathering material, which may include grass as well as seaweed. Both members of the pair participate in building.

The normal Kittiwake clutch is of two eggs, but occasionally one or three are laid. The eggs are dull grey-green with dark spotting, as is usual

Nesting material includes pulled-up grass as well as seaweed gathered on the shoreline.

The young birds' first set of feathers bears a bold pattern in black, white and grey, quite unlike the streaky brown juvenile plumage of most gulls.

for gulls, and take 25–32 days to hatch, the parents sharing incubation duties. The first-laid egg hatches a day or two before the second. The chicks have silvery-white down when hatched, quite different from the brown-speckled down of our other nesting gulls – this reflects the fact that they have little need for camouflage. Instead they are reliant on their inaccessible cliffside nests, and the collective defensive force of the adults in the colony, should a skua or large gull come drifting past in the hope of snatching a chick from the ledges.

The parents feed the chicks on regurgitated fish, and brood them for the first few days of their lives. One adult often stays in attendance after the chicks no longer need to be brooded, but as the chicks grow they need to be brought food by both parents – at this stage the youngest chick in a brood of three is quite likely to succumb to starvation. The chicks fledge at 33–54 days and rapidly gain the skills they need for independent life. Those that survive their first year stand a good chance of living into double figures; the longevity record is 28 years and six months.

MOVEMENTS AND MIGRATION

Kittiwakes move out to the open sea after breeding, with most birds departing in July or August, and generally heading south-west, with concentrations around the Bay of Biscay. Many older birds do not travel very far from where they breed, but young birds are more adventurous and a significant proportion cross the Atlantic to winter off Greenland, Newfoundland or the north-eastern USA. Others move south – ringed birds have been recovered in Algeria and Western Sahara, as well as in the Mediterranean off Italy, and round north-eastern Europe to Finland and the edge of western Russia. A few recoveries have been made thousands of kilometres inland.

In winter, adult Kittiwakes develop a dark marking behind the eye and a greyish 'neck boa'.

First-winter birds, sometimes nicknamed 'tarrocks', wander the open sea for their first few months of life.

THE FUTURE

The population trends of the Kittiwake over the last century are similar to those for other seabirds such as the Arctic Skua. Persecution into the 20th century kept numbers low, but its cessation allowed the population to grow through the later decades, peaking in around the mid-1980s. However, this overall trend masked the fact that some colonies in southern and western Britain and southern Ireland were already starting to shrink in the 1970s, and over the next couple of decades the large North Sea colonies began to suffer declines as well. Between Operation Seafarer in 1969–1970 and the Seabird Colony Register in 1985–1988, Britain and Ireland's population increased by 24 per cent, but between the Seabird Colony Register and Seabird 2000 there was a decline of 25 per cent, with a further decline of 47 per cent in 2000–2012.

Kittiwakes were historically killed for their wings, which were used as decorations for hats, a practice that was much reduced after the introduction of the Sea Birds Preservation Act in 1868, although it did not really cease until around the time of the First World War. They were also persecuted because they prey on fish, a resource that humans use heavily, but thankfully they now enjoy full legal protection. Nowadays, it is possible that changes in fish distribution and abundance, perhaps connected to global warming, pose the most serious threat to the Kittiwake population. In any case, a dearth of suitable fish at the right time is making it more difficult for Kittiwakes to provision their chicks, and in 2000–2012 there have been a few years where breeding success over large areas of the Kittiwake's range has been incredibly low.

Although Kittiwakes prey on a range of fish species, it is shortages of immature sandeels in particular that are correlated with low breeding success. Other seabirds are also affected, but the impact is often worse for Kittiwakes because they have less flexibility in their foraging behaviour. They are not able to dive deeply (like auks), nor are they adapted to make very long foraging flights when necessary (like shearwaters). Adult survival over winter has remained generally high, suggesting that breeding failure rather than the loss of too many adult birds is driving the falling numbers. When sandeel numbers are low, parent Kittiwakes turn to other fish species, which may be quite unsuitable. For example, observations on Fair Isle in 2006 and 2007 revealed that the adults were bringing large numbers of pipefish to the nests – weight for weight, these tough-bodied fish offer a fraction of the nutritional value of sandeels. Fair Isle's Kittiwakes have suffered a string of disastrous breeding seasons, with no chicks fledged at all in either 2011 or 2012.

There are a few other potential causes of Kittiwake declines, which include increased predation by Great Skuas in some Scottish localities, and poisoning from toxin-producing algal blooms, known as 'red tides'. The latter caused high mortality in north-east England in 1996–1997 and may have been involved in some localised declines in the 1980s. Kittiwakes are sometimes affected by oil spills, although not to the same extent as auks. Also, some colonies on buildings have been threatened by redevelopment.

To halt and reverse the decline in Kittiwakes, a fuller understanding of sandeel biology and distribution is necessary, so that key foraging areas can be protected from overfishing. However, it must be borne in mind that there could be natural fluctuations at work as well.

Mediterranean Gull *Ichthyaetus melanocephalus*

A relatively new addition to the list of British breeding birds, since the 1950s this beautiful gull has spread dramatically from its original stronghold in Ukraine and elsewhere around the Black Sea, to colonise much of western Europe. It has been breeding in Britain since 1968 and can now be seen at many sites in England, and more widely in winter.

INTRODUCTION

This gull looks superficially like a large Black-headed Gull but has a heavier, bright red bill, pure white primaries, and in breeding plumage a blacker 'hood' with a striking white partial eye-ring. Immature birds are best identified by their size and bill shape, and the prevalence of dark grey rather than brown in the retained juvenile feathers in the wings. The call is a very distinctive querying 'yowk'.

DISTRIBUTION, POPULATION AND HABITAT

Colonisation of Britain began in south-east Kent and this area is perhaps still the species' British 'base', but breeding birds can now be found along the south coast into Dorset, and in north Kent and East Anglia. There are also breeding birds in north-west England. In winter the species is seen regularly all around the coasts of England, Wales and much of Scotland and Ireland, and also at some inland areas, especially in the Midlands. There are 600–630 pairs breeding here and the wintering population is 1,800 individuals. Worldwide, the breeding population is somewhere around 184,000 pairs, most of which are still found on the Ukrainian side of the Black Sea.

Most Mediterranean Gulls in Britain nest among Black-headed Gull colonies.

The Mediterranean Gull breeds on lightly vegetated islands and the edges of coastal lagoons, and on saltmarshes. It often shares its habitat with nesting Black-headed Gulls. Non-breeding birds are mainly found on coastlines with beaches, but they do also join other gull flocks inland to roost on reservoirs and feed on rubbish dumps.

BEHAVIOUR AND DIET

This gull is gregarious at all times, and is usually seen with others of its own kind and/or other gull species. It has a relaxed and leisurely wheeling flight and a strutting walk, and at colonies it dominates the more numerous Black-headed Gulls. On the breeding grounds it eats mainly insects, which it picks from the water or just above it, and at other times it scavenges for carrion and shellfish on the shoreline. Inland it may follow the plough in search of worms in the turned-over soil, and visits rubbish dumps to scavenge for all manner of scraps.

BREEDING

In newer and smaller colonies, Mediterranean Gulls may form hybrid pairs with Black-headed Gulls, but this is rarer at more established sites. However, most colonies are within or alongside larger colonies of Black-headed Gulls. Within the colony Mediterranean Gull pairs often commandeer the favoured central positions. They dig out a shallow nest scrape and in May the clutch (usually of three eggs) is laid. Both parents share in incubation duties and the rearing of the chicks – incubation takes 24 days, and the chicks fledge in another 35–40 days. Before fledging they wander some distance from the nest and may even enter nearby water and swim to other parts of the colony. They are vulnerable to both land- and air-based predators, but the entire colony reacts when danger threatens and is often able to repel a predator. By mid-July most adults and chicks have left the colony.

Many of the Mediterranean Gulls in Britain over the winter are visitors from eastern Europe.

MOVEMENTS AND MIGRATION

Adult Mediterranean Gulls do not move great distances individually as a rule, although there is a general westwards push in terms of colonising new ground. Young birds go further – there have been recoveries and sightings of British-bred young birds in Spain and the Azores. Several European projects on Mediterranean Gull movements are underway, whereby the birds are uniquely marked with colour rings that can be read from a distance. These studies have shown that many of our wintering Mediterranean Gulls come from eastern Europe, in particular Germany, Hungary and Poland, with others arriving from Belgium, the Netherlands and France. Several of our breeding birds move across the Channel to northern France, the Netherlands and Belgium in the winter.

THE FUTURE

This gull seems to be making its own fortune, with its dramatic spread and population increase in Britain. However, it is still, in both global and national terms, a rare bird, and its population as a whole is not increasing. In eastern Europe it faces various threats, including destructive developments at its breeding grounds, and harvesting of both eggs and adults from colonies around the Mediterranean.

The species' success as a British breeding bird is due at least in part to the fact that most colonies are on nature reserves, so the habitat is both protected from development and managed to benefit wildlife. If it takes to nesting inland within Black-headed Gull colonies there is much additional habitat available to it, but so far it has seemed disinclined to breed away from the coast.

Black-headed Gull *Chroicocephalus ridibundus*

Whether sporting its chocolate-brown hood or wearing its whiter winter plumage, this small gull is a dapper and characterful bird. It is also a very familiar species due to its abundance in urban environments, both coastal and inland. It has a very broad distribution in the British Isles, and its coastal colonies in particular are often shared with other ground-nesting seabird species. It is our only widespread gull species to have shown a clear population increase since the turn of the century; its adaptable habits seem to be invaluable in driving its success here. However, on a global scale its situation is less encouraging, with evidence that numbers are falling in parts of its range.

INTRODUCTION

The Black-headed Gull is the only dark-hooded gull likely to be seen in spring and summer over most of the British Isles. The most likely confusion species is the much scarcer Mediterranean Gull, but the two are easy enough to separate given good views. The Black-headed Gull's hood is brown rather than black and stops quite high up on the nape compared with the fuller, jet-black hood of the Mediterranean Gull. There is a partial white eye-ring, less prominent than that of the Mediterranean Gull. The hood is lost in autumn, reappearing in early spring (although the timing varies from bird to bird); in winter just a small dark smudge behind the eye remains on the otherwise white head. In winter adults have bright red bills, but in the breeding season the bill is darker. Adults also show a distinctive wing pattern – the primaries have very small black tips and the outermost primaries are otherwise white, creating a long and obvious white leading edge to the 'hand' of the wing.

The dark 'hood' of breeding plumage develops any time from very late autumn to early spring.

The pattern of the adults is present in first-winter Black-headed Gulls as well. This species is a two-year gull, with a very short-lived, mostly sandy-brown juvenile plumage. This is quickly replaced by the very adult-like first-winter plumage, although first-winter birds do show some brown in the wings and a dark tail-band. The gull is light and buoyant on the wing and manoeuvres with agility. It sits high in the water when swimming, and when perched looks quite tall and leggy.

Black-headed Gulls usually lose their brown juvenile plumage before they leave the breeding colony.

DISTRIBUTION, POPULATION AND HABITAT

This gull can be seen almost everywhere in Britain and Ireland, at all times of the year, although it is generally most widespread and much more numerous in winter. There are particularly large concentrations of breeding birds in Northern Ireland, on the Hampshire coast and adjoining counties, on the coasts of Essex, Suffolk and north Norfolk, in Lancashire and Merseyside, and in eastern Scotland up to Orkney and Shetland. There are about 140,000 pairs in the UK and another 3,900 in the Republic of Ireland, amounting to about 6 per cent of the world population, while our wintering population may reach more than 2 million individuals.

This gull breeds across central Eurasia in a broad band, avoiding both the far north and the far south. It has also recently begun to colonise the east coast of North America, although numbers breeding there are still low. Its total world population is estimated to be 4.8–8.9 million individuals.

Many habitat types can be suitable for nesting Black-headed Gulls. Some breed on shingle islands and artificial floating structures within gravel workings, both in coastal areas and inland. They also nest on saltmarshes, wet grassland and moorland, and on sand dunes. They feed at sea, on beaches, on grassland, in agricultural fields, in town parks and on rubbish dumps.

BEHAVIOUR AND DIET

Black-headed Gulls are highly social, forming close-knit colonies and living in flocks in the winter months. They readily flock with any other gull species, and sometimes also with terns. They are lively and very vocal birds, often calling to each other even in winter with a harsh and shrill 'kee-yaaah', while the noise that emanates from a large breeding colony in full swing can be deafening.

On the ground they have a strutting walk, and on the water they float buoyantly with their wings tilted well clear of the surface. They spend much time preening and bathing, and when engaged in the latter often take off and rise a metre or two before plunging back into the water. They may amuse

A CONFUSING STRING OF NAMES

It is one of the curious conundrums in bird nomenclature that the Black-headed Gull is so called, when its head is actually brown. The Mediterranean Gull's scientific species name, *melanocephalus*, actually means 'black-headed' and the name is a better fit for this species, which does have a genuinely black head. The Black-headed Gull's species name, *ridibundus*, means 'laughing', and alludes to the bird's call, but this name is not ideal either as there is a species called Laughing Gull in North America. The string of misplaced names continues – the Laughing Gull's species name is *atricilla*, meaning 'black-tailed', and another American species exists with the English name Black-tailed Gull. This bird is the final link in the chain – its species name is *crassirostris*, meaning 'thick-billed', but there is no species with the English name Thick-billed Gull.

Ploughing exposes plenty of worms and leatherjackets, valuable winter food for Black-headed Gulls.

themselves by picking up leaves from the surface and dropping them again, an apparent form of play, but also useful practice for catching food in flight. They are quite skilled at catching flying insects, and their prowess at catching flying crusts of bread thrown from park bridges is well known to many people who visit their local park to feed the birds.

A Black-headed Gull 'mugs' a Puffin to try to steal its catch of sandeels.

Food types taken by these gulls are many and varied. They may catch small fish and other swimming organisms at sea and on fresh water, although on the latter they are more likely to catch aquatic insects. They also eat carrion of every kind and take eggs from other birds' nests. In rural areas they follow the plough, while in towns they pick up dropped food from the street. They also visit refuse tips to search for anything edible.

BREEDING

This highly colonial bird returns to its breeding grounds, sometimes shared with terns, in March or April, and the males attempt to establish and defend a territory. Competition is fierce for the best spots, which are towards the centre of the occupied area. Once a pair has formed, which is likely to be the same partnership as in previous years, both birds take part in nest building, using whatever vegetation is found locally to construct a basic nest that is little more than a pile of leaves and stems. Occasionally the nest may be off the ground, in a bush or tree, or on a building. Pairs are territorial around their nests, even though the next pair along may be just a few centimetres away, and often noisy arguments break out. However, the colony responds as one when a potential predator, such as a large gull, hoves into view, rising to mob and chase the intruder until it moves on. The largest colonies are thousands of pairs strong and present a real force to be reckoned with.

The clutch is usually of three eggs, which are incubated by both sexes for 23–26 days. There is a small age difference between the chicks when they hatch. They wander out of the nest to explore the surrounding area when they are about ten days old, and if there is open water nearby they readily go to the water's edge to paddle or swim. They are tolerated by their neighbours as they make their way through other territories to the shoreline, and sizeable crèches of part-grown chicks may form on the water. The adults continue to feed them regularly, regurgitating their crop contents in response to the chicks' begging calls. By 35 days old the chicks have replaced their down with the gingery juvenile plumage, and are able to fly. They become independent very soon afterwards.

These gulls nest in close-knit colonies, using strength of numbers to keep predators away.

MOVEMENTS AND MIGRATION

Young Black-headed Gulls head off in random directions from their natal colony – chicks ringed in Surrey in a large-scale project were sighted within weeks in places as diverse as Ireland, Wales and France, as well as widely separated sites in England. After the initial dispersal they are likely to head south in winter, mainly to the coasts of France and Spain, but a few make quite epic journeys, with one Cumbrian chick travelling to Senegal and two other English-bred chicks reaching Mauritania. Established breeding birds based in England generally overwinter quite close to their breeding grounds. The many extra birds that visit Britain and Ireland in winter come mainly from Scandinavia and Russia, with a few from Iceland.

THE FUTURE

Black-headed Gulls are regarded as highly successful and sometimes a nuisance, threatening the breeding success of other, scarcer bird species that share their habitat. However, although their population has generally increased in the late 20th and early 21st centuries, there have been some significant local declines, especially in Ireland and Scotland. Possible causes of these include agricultural change and predation by American Mink. There is also evidence that the breeding population is gradually becoming more concentrated at fewer sites, with smaller colonies disappearing. It is important to keep a close eye on Black-headed Gull numbers, both in general and at specific key colonies, because it cannot be assumed that the species is immune from the factors that have brought about declines in our other gull species.

Common Gull *Larus canus*

Of all our gulls, this species is perhaps the one with the most tenuous association with the sea, as much of the population breeds and winters inland. However, it can still be seen on the coast in varying numbers and does have some significant coastal colonies as well as many that are inland. It is an attractive medium-sized gull with a gentle expression and a natural elegance that sets it apart from its larger relatives.

INTRODUCTION

The adult Common Gull looks rather like a Kittiwake, with its mid-grey wings, all-yellow bill and large dark eyes that give it its pleasantly mild expression, but its shape is more like that of a small Lesser Black-backed Gull, with quite long legs and an upright stance. It has greyish-green or yellowish legs, and in winter develops fine streaking on its head and a smudgy black ring near the tip of its bill.

Juveniles are quite unlike young Kittiwakes and have a similar mottled grey-brown colour scheme to juveniles of the larger gulls, but they acquire adult plumage over three rather than four years. In flight this gull looks light and buoyant, with long wings and a slim outline.

DISTRIBUTION, POPULATION AND HABITAT

The breeding population of this bird in the British Isles is concentrated in Scotland and northern and western Ireland. Orkney, Shetland and the north-east of the Scottish mainland have particularly large numbers. Away from these core areas there are just a handful of small colonies. The total breeding population is about 49,000 pairs. In winter Common Gulls from further afield arrive here and numbers swell to about 710,000. They can then be found almost everywhere in the British Isles, only avoiding the most unsuitable habitat types.

Although superficially like a Kittiwake, the Common Gull differs in having white in its wingtips, and greenish, much longer legs.

NOT SO COMMON?

Many beginner birdwatchers, especially those living in England and Wales, are perplexed that this gull species should be called the Common Gull when others, such as Herring and Black-headed Gulls, seem much more numerous. However, in this context 'common' does not mean 'abundant', but refers to common land, describing the species' fondness for well-grazed grassy fields as foraging ground.

Worldwide, the Common Gull is distributed across northern Eurasia (three subspecies, *canus* in the west, *heinei* in Central Asia and *kamschatschensis* in the east) and northern North America (the subspecies *brachyrhynchus*, known as the Mew Gull and sometimes split as a separate species). The world population is estimated to be 2.5–3.7 million individuals.

This gull breeds on the banks of lochs and rivers and on islands within them, and on wet moorland, marshes and grassland with small pools; it occasionally nests in small trees and in some areas on buildings. In winter it can be seen on arable fields, around gravel pits and reservoirs, in town parks with lakes and on beaches.

Far more Common Gulls are in Britain in winter than in the breeding season.

BEHAVIOUR AND DIET

The Common Gull is a gregarious bird, flocking with its own kind and other gulls (especially Black-headed Gulls) in winter, and breeding colonially. It is quite an aerial bird, often patrolling around fields or along shorelines on the wing, in search of feeding opportunities. It may pursue aquatic prey, including small fish and, on fresh water, insect larvae, by dip feeding in flight or by splash diving. It swims comfortably, sitting very buoyantly on the water.

Common Gulls are frequently seen on fields, walking steadily along and checking the ground for invertebrate prey such as earthworms and leatherjackets. They 'dance' on the grass to simulate rainfall and draw worms to the surface (see page 142). They also follow the plough, and on beaches they forage among rock pools for small shrimps, crabs and molluscs. When there is a hatch of flying ants they take to the air and pursue the insects with great agility, and they take birds' eggs and chicks if the opportunity presents itself. Their diet also includes carrion and scraps from rubbish dumps. In town parks they take handouts of bread meant for ducks and pigeons.

Young chicks have a warm coat of down, and are able to potter around at a couple of days old.

BREEDING

Some pairs of Common Gulls nest alone, but they are much less likely to be successful than those that nest communally – a mostly white bird sitting on a ground nest is very conspicuous to predators, and a single pair has little chance of chasing away a determined crow or other mid-sized predator. Colonies may be several hundred pairs strong – colony size depends on the area available for nesting – and all pairs within a colony benefit from having their neighbours' help with mobbing any would-be predators. Some colonies are mixed, also holding pairs of Black-headed Gulls or Herring Gulls.

Each pair holds a small territory within the colony, and is usually loyal to that site year on year. The territories near the centre of the colony are most likely to be successful, so are occupied by the older, more experienced birds, while first-time breeders are forced to take one at the colony edge. Males are first to occupy their territories, arriving in March, and call to attract a female, while females that are established breeders head for their previous year's nest-site, where they may well find their previous year's partner already in residence. Typical gull pair-bonding behaviours then ensue, leading up to much courtship feeding in the last week or so before eggs are laid. Pairs bond for life in most cases, although separation and relocation can occur if a pair has a failed breeding attempt.

RAIN DANCE

The sight of a group of gulls performing an energetic Riverdance-style routine on a football field is amusing enough to catch the attention of birders and non-birders alike. The gulls stand on one spot, then rapidly drum their feet on the ground, before pausing to scrutinise the grass around them. They may then move on to a different spot before repeating the routine. This curious and comical behaviour is performed to create the illusion of heavy rain falling on the ground, a stimulus that has a dramatic effect on earthworms. To avoid being flooded out of their tunnels, the worms head towards the surface, where the waiting gulls seize them. Similar tactics are used by some of the (human) participants at the worm-charming contest that takes place each year in Caerphilly, Wales. Some very skilled worm-charmers can draw 500 or more worms from the ground in just half an hour, so the method is probably quite an effective, if tiring, way for a hungry gull to obtain a meal.

The nest is constructed from various kinds of local vegetation, and is usually on the ground but may be in a tree; in a few locations (most notably in Aberdeen) Common Gulls are acquiring the rooftop-nesting habit. The eggs are laid in May and the clutch usually consists of three eggs. They are greenish-brown and spotted, and are laid at one- or two-day intervals. Incubation begins when the first egg is laid so that the clutch hatches asynchronously. The parents share duties during the 22–28-day incubation period. Soon after hatching the downy chicks are quite mobile and may wander from the nest if space allows, but will not stray very far. They develop quickly on a diet of regurgitated insects and other prey, and can fly after 35 days, becoming fully independent soon afterwards.

This gull nests on the ground, where well-grown vegetation can help hide the chicks from danger.

MOVEMENTS AND MIGRATION

Adults and young birds leave their breeding grounds by August and many move south and west, to winter either inland or on the coast, while some head for the North Sea coast. Some young birds leave Britain and head down the Atlantic coast of France and Spain, or into the Mediterranean. Many of the Common Gulls caught and ringed in Britain in the winter have subsequently been recovered or spotted at breeding colonies in Denmark, Sweden and Norway, and a few more in Russia.

THE FUTURE

Coastal-breeding Common Gulls have increased in number historically, with a rise of 25 per cent between Operation Seafarer in 1969–1970 and the Seabird Colony Register in the 1980s, and another increase of 36 per cent between the Seabird Colony Register and Seabird 2000, although this may partly reflect improved coverage in the later surveys. Operation Seafarer and the Seabird Colony Register did not count inland colonies, but there is evidence from other surveys that some of these have shown quite severe declines. A colony on the Correen Hills in Moray went from 24,500 nests in 1988–1989 to none at all by 1998. The population could have simply relocated, but overall inland colonies appear to be declining. Farming and land-management practices could potentially have a significant impact on Common Gulls nesting inland. Drainage of wet grassland, and conversion of moorland to pine plantation, both deprive the gulls of breeding habitat.

The picture of coastal colonies since Seabird 2000 is unclear, with some declining and others increasing. In some parts of Scotland predation by the non-native American Mink has had a severe impact on the breeding success of coastal breeders, lowering productivity by up to 76 per cent. Mink trapping is necessary in these areas, to help the Common Gulls and many other ground-nesting birds.

Outside Scotland there are very few Common Gull colonies. In England and on the Isle of Man, Seabird 2000 found only 39 nests, and none at all in Wales. Surveys indicated that little had changed in England in 2012, and Wales has not held any breeding Common Gulls since Operation Seafarer, which found the grand total of two nests. The total for the whole of Ireland was 969 nests – this actually represented a sizeable increase from the Seabird Colony Register, which found 301 nests. Unfortunately very little data has been collected from 2000 onwards. Any future increases in Common Gull populations outside Scotland will probably depend on how the Scottish colonies and also those on nearby mainland Europe fare over the coming years.

Unlike the larger gulls, Common Gulls in their first winter show almost pure white undersides.

Lesser Black-backed Gull *Larus fuscus*

A fairly large but elegantly built gull, the Lesser Black-backed Gull is an abundant but – at present – declining breeding bird in Britain. It is often confused with other gull species, particularly in subadult and non-breeding plumages, and the picture is confused by the existence of several subspecies. Only recently has the Lesser Black-backed Gull begun to overwinter in the British Isles in large numbers, and it has also become more numerous as an urban breeding bird over recent years.

INTRODUCTION

This gull is part of the 'large white-headed gull' complex that forms the core of the genus *Larus*. It is usually slightly smaller and looks distinctly more slimline and long winged than the Herring Gull, and in breeding plumage has bright yellow legs, a clean white head and underside, and a dark grey mantle and wings that are visibly lighter coloured than the black, white-spotted wingtips. In winter the head develops extensive brownish streaking and the leg colour becomes duller. When a pair is together it is usually possible to tell which bird is which, as males are a little larger than females and have a more angular head shape.

Our breeding subspecies, *graellsii*, is relatively pale on the mantle and wings, but in winter some of the birds that visit are of the darker subspecies *intermedius*, which is difficult to tell from the Great Black-backed Gull by back and wing colour alone. An even darker subspecies, the nominate *fuscus*, is a very rare visitor. In juvenile plumage the gull is mid grey-brown with darker streaks and chequering, and as it matures over four years it gradually assumes the adult plumage pattern and colours.

This is a strongly built gull, but a little smaller and slimmer than the usually more common Herring Gull.

In winter, Lesser Black-backed Gulls range very widely across Britain and young birds in particular often move south to mainland Europe.

DISTRIBUTION, POPULATION AND HABITAT

Lesser Black-backed Gulls can be found all year round along almost the entire coastline of Britain and Ireland. The largest colonies on the coast are in eastern Scotland around the Firth of Forth, and in western Scotland, Cumbria, Lancashire, Pembrokeshire, East Anglia and the Isles of Scilly. More than half of the UK's breeding population is concentrated at just ten sites, with Walney Island in Cumbria holding the single biggest colony. Breeding colonies are smaller and more scarce along the south coast, in between Norfolk and Northumberland, and in north-east Scotland (although there are strong populations on Orkney and Shetland).

GULL IN THE RING

The Lesser Black-backed Gull, the Herring Gull and a few related species found in Eurasia and North America are frequently cited to form an example of a ring species. This is an important concept in evolutionary biology, and describes a series of populations of animals that form a looping chain geographically, and freely interbreed with their immediate neighbours but not with those further along in the chain. While neighbouring populations are very similar and form a continuum, those at either end of the chain are very distinct from each other, and do not interbreed even though they are found in the same places. In the case of the gulls, the ring circles around the North Pole, and the populations (moving eastwards from Britain) are: **1.** Lesser Black-backed Gull (subspecies *graellsii/intermedius*), **2.** Lesser Black-backed Gull (subspecies *fuscus*), **3.** Heuglin's Gull *Larus heuglini* (sometimes classified as a subspecies of Lesser Black-backed Gull), **4.** Vega Gull *L. vegae birulai*, **5.** Vega Gull *L. v. vegae*, **6.** American Herring Gull *L. smithsonianus* and finally back to Britain and the **7.** Herring Gull, which lives alongside the Lesser Black-backed Gull but does not normally interbreed with it.

A genetic study in 2004, however, indicated that this representation is in fact an oversimplification, and does not account for several other gull species which, in terms of relatedness, would be a part of the same 'superspecies'.

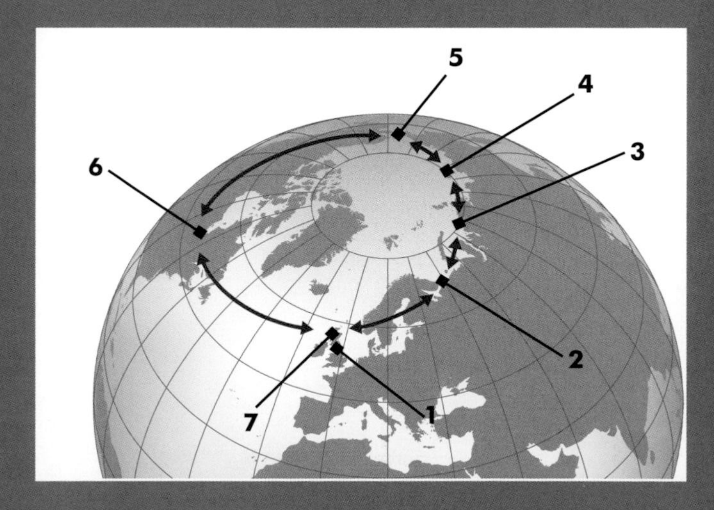

In Britain and Ireland the Seabird 2000 survey indicated that there was a total of some 115,000 pairs, 110,000 of them in the UK. There is, however, evidence that there has been a steep decline since those figures were recorded. The British and Irish population represents nearly 40 per cent of the world's total, and more than 60 per cent of the subspecies *graellsii*. This subspecies is also found in France, Spain, Portugal, the Faroes, Greenland and Iceland. The winter population is about 130,000 individuals, a sizeable proportion of which are *intermedius* birds from Scandinavia, while many of our own breeding birds migrate south for the winter. The subspecies *intermedius* breeds mainly in Norway and Sweden, while *fuscus* is present as well in parts of those countries but also occurs further east, in Finland and Estonia.

Lesser Black-backed Gulls nest in various habitats, from islands on lakes to clifftops, moorland, dunes and the tops of buildings. They often breed in mixed colonies with Herring Gulls. In the winter they range quite widely, flocking on fields to hunt for worms and other invertebrates, visiting town parks to join ducks and geese in the squabble for handouts of bread, and assembling on rubbish dumps to scavenge for scraps. They join large mixed-species roosts on gravel pits and reservoirs.

BEHAVIOUR AND DIET

These gulls are highly gregarious, especially in winter, and form a significant proportion of most mixed-species gull gatherings. In winter they roost communally on water, as this offers protection from most predators, and at dusk and dawn they may be seen commuting to and from the roost sites, flying in small or medium-sized flocks, sometimes in a V formation. In urban areas they can become very bold, and in city parks and seaside towns they may even approach people in the hope of being given food. They are noisy birds, like most gulls, with various harsh cackling and cawing calls.

Although we most often see this gull when it is scavenging for food on shorelines or among rubbish, it is a competent hunter of live prey and, particularly in the breeding season, catches fish well out at sea. It may pick prey from the surface while in flight, or make a shallow splash dive. Its feeding range at sea has been found to be more extensive than that of other large gull species.

Besides preying on live fish, the species also preys on crustaceans found in both deep and shallow water as well as on the shore. It eats molluscs and may drop them from a height to break their shells. It is very much an opportunist and does not hesitate to attack and kill an injured bird in order to eat it, raid other birds' nests (including nests of its own species), catch a small mammal near its nest if it gets

A few Lesser Black-backed Gulls join a large mixed gull flock to feed on razor shells washed up on a Norfolk beach.

the chance or chase another gull to try to steal food from it. Flocks of Lesser Black-backed Gulls follow fishing vessels to pick up fish discards, and tideline carrion is an important food source. In towns the birds also take roadkill and edible litter, and in the wider countryside they feed on invertebrates, especially earthworms, exposed by the plough, as well as all manner of foodstuffs found on rubbish tips. The diet also includes some natural plant materials such as berries and seaweed. Like all gulls the Lesser Black-backed Gull regurgitates pellets composed of the indigestible parts of prey, and sometimes other indigestible material that was swallowed in error.

BREEDING

Adult Lesser Black-backed Gulls return to their breeding grounds as early as December, allowing plenty of time for courtship, nest building or repair, and building body condition for breeding before the eggs are laid in April or May. They are usually faithful to one partner for life and also stick to the same nest-site if they can. Separation may occur, however, in the year following a breeding attempt that was unsuccessful. Hybrid pairs between Lesser Black-backed and Herring Gulls form on occasion, and such pairs can successfully produce chicks.

Territorial behaviour includes much vocalisation, including the drawn-out 'long-call', of which all our large gulls have a slightly different version.

In some years a high proportion of pairs do not attempt to breed at all. This is an expected strategy for a long-lived species to contend with times of poor physical condition – and this is indeed a long-lived bird, with several ringing recoveries of birds in their thirties. Breeding failure may result in 'divorce' of the pair but, as newly formed pairs are likely to have lowered breeding success for more than one season compared with established pairs, divorce is not necessarily the best option or the one chosen by the pair.

The period of pair formation and bonding is characterised by various calls and posturing between the birds, including a rapid parallel walk with the head tilted downwards. The pair also shows aggression towards intruders that might attempt to usurp the nest-site. Although they breed in colonies, the birds are territorial and both sexes will attempt to see off others of their species over a territory that may cover some 5 square metres. In the run-up to egg laying, the male gull feeds his mate frequently. The female solicits this by approaching the male in a low, hunched posture, and making a distinctive 'mew' call while tossing her head – the same behaviour can be seen in well-grown chicks begging their parents for food. The head-tossing display, performed by both birds, is also a prelude to copulation.

STRANGER DANGER

The advantages of colonial life are clear – all birds benefit from extra eyes and ears to detect approaching predators, and extra strike force when driving those predators away. One significant disadvantage comes when your neighbours in the colony are predators themselves, and may consider your offspring to be fair game. A study of Lesser Black-backed Gulls nesting on Skokholm island, off Pembrokeshire, found that pairs which nested late in the season were fairly likely to lose their brood, most often at egg stage, to another Lesser Black-backed Gull. Nests in the densest part of the colony were also most likely to be predated, and pairs that had lost their own eggs or chicks were the ones most likely to help themselves to their neighbour's nest.

The nest is positioned centrally within a territory and is a large pile of mainly plant matter. If built on the ground it is often located close to some cover in which the chicks can take refuge if need be when the time comes. Both members of the pair bring in nesting material and shape the nest, and in late spring the female lays her clutch of usually three eggs (sometimes two or four). Both parents share incubation duties, which take 24–27 days. As incubation commences when the first egg is laid, the hatch is asynchronous, with the oldest chick holding a clear advantage over its younger siblings when competing for food from the adults. Both eggs and chicks are well camouflaged with a brownish ground colour and darker mottling. The downy chicks are mobile soon after hatching, and leave the nest when a few days old to find separate hiding places nearby (although their options may be limited, depending on the nest-site – in the case of rooftop nests, for example, the chicks may stay in the nest for most of the pre-fledging period because they cannot reach any other safe areas).

This gull may nest on cliffs but more often on flat or sloping vegetated ground, which gives the young chicks more room to roam.

As they grow the chicks shed their down to reveal a coat of fresh juvenile feathers, which are grey with darker spots and streaks. By the age of three weeks they are very active and spend much time exercising their wings. When a parent arrives with food, the chicks hurry over to the adult, giving loud and plaintive whistling calls, and beg by pecking at the red spot on the adult's bill. This stimulates it to regurgitate a crop-load of partly digested food, which the chicks squabble over. The young birds can fly at about 35 days old, and for the next few weeks follow their parents relentlessly, begging to be fed, but gradually learning how to find their own food. By August all birds disperse from the breeding grounds.

After breeding, some Lesser Black-backed Gulls move to parkland with lakes, a suitable habitat for the winter months.

MOVEMENTS AND MIGRATION

Many British-breeding Lesser Black-backed Gulls migrate south for the winter, but most do not travel as far as they did historically, and many stay very near the breeding grounds. This may be an adaptation to exploit increased supplies of food available to them in the winter, especially at refuse tips. Young birds are most likely to roam and a fair proportion of them go south to France, Spain or North Africa, wintering on the Atlantic or western Mediterranean coastline. There has been a scattering of recoveries from well inland on the African continent, most remarkably one in Chad. There are also many recoveries of British-bred birds in Iceland and Scandinavia.

Birds of the *intermedius* subspecies begin to arrive in Britain from October, and stay until mid-spring. Several ringed *intermedius* birds from Norway and Sweden have been found in the British Isles.

THE FUTURE

The recent decline of the Lesser Black-backed Gull in Britain is perplexing. The picture throughout the 20th century is one of steady increase, as the birds bounced back from many decades of intense persecution and, in some areas, large-scale harvesting of their eggs. At the same time new feeding opportunities were opening up for the birds on the coast in the form of new fisheries and the waste they generated. Inland, things seemed even better. Ever more sizeable landfill sites provided ample food, and town rooftops made ideal nesting sites. Numbers of roof-nesting pairs quadrupled in 1993–2002 across Britain and Ireland, while coastal pairs in Britain and Ireland rose from 50,035 in the 1969–1970 Operation Seafarer survey to 64,417 in the Seabird Colony Register (1985–1988), and had reached 91,323 for Seabird 2000.

Since Seabird 2000, the population in the UK at least is believed to have fallen by a startling 50 per cent, although this is extrapolated from counts taken from selected colonies as the next full seabird census has yet to be completed. Data from the colony on Skomer suggest that reduced adult survival is behind the decline that has been recorded there, but what is causing the adult birds to die in higher numbers is unknown. Food supplies in the form of fishery discards are certainly now falling, and this could be affecting winter survival. Many inland colonies still seem to be thriving, and indeed Lesser Black-backed Gulls can legally be killed by any landowner under the general licence, if they are considered to be causing a health hazard. Inland colonies, however, make up a small minority of our total breeding population, and investigation into possible causes behind the declines of coastal birds is urgently needed.

The subspecies *intermedius*, which visits Britain in winter, is noticeably darker-backed than our breeding *graellsii* birds.

Herring Gull
Larus argentatus

With its mewing and yelping calls, graceful flight, silver wings and feisty, assertive personality, this gull embodies the spirit of a British seaside town more than any other. Loved and hated in equal measure by people who live by the sea, the Herring Gull is probably the most familiar of all our seabirds, and in the winter it is also a familiar land bird that can be seen almost everywhere in Britain and Ireland. Yet despite this apparent great abundance it is a species in trouble, experiencing a long-term decline that has led to it being assigned a Red status of conservation concern in the UK.

INTRODUCTION

The adult Herring Gull has the classic 'seagull' colour scheme of white head and body, light silver-grey wings and black wingtips marked with white spots. The bill is yellow with a red spot near the tip of the lower mandible, the legs and feet are pink, and the eyes are a light straw colour, set in a narrow ring of bare yellow skin. The pale eye gives the bird a rather harsh facial expression, in common with other large white-headed gull species. In winter the head and neck have a variable amount of brown streaking.

Adult plumage is acquired over the first four years of the bird's life. The mottled brown juvenile plumage, with black bill and dark brown eyes, is gradually replaced with adult colours. Birds in their third winter look very adult-like but show a little brown in the wings and tail-tip, and some dark markings on the bill. The Herring Gull is a powerfully built bird, larger and stockier than the Lesser Black-backed Gull and closer to a small Great Black-backed Gull in build. Females are smaller than males, with more rounded and 'gentler-looking' heads.

This well-known gull has a relatively short-bodied, strong-billed outline, with pink legs and a mid-grey back and wings. Its head is pure white in summer but develops prominent dusky streaking in winter.

The classic seaside gull, this species is a familiar sight over coastal towns and villages throughout Britain.

DISTRIBUTION, POPULATION AND HABITAT

Herring Gulls breed around almost our entire coastline, although there are some gaps along the very flat coastlines of South Yorkshire, Lincolnshire and north-east East Anglia, and parts of south-west and western Ireland. The largest colonies are in Scotland, especially on the west coast, and west Wales. Most of the colonies along the south coast are on buildings, as are many in south-west and north-east England. The Seabird 2000 census counted 147,114 nests across Britain and Ireland, but the figure is probably now considerably lower. Numbers in winter soar to some 740,000 individuals, many of them visitors from Scandinavia, and the birds can be found across most of Britain and Ireland, only avoiding mountainous areas and those with forest cover.

Herring Gulls have an extensive distribution across northern Europe and perhaps beyond, although the picture is somewhat muddled by taxonomy, with several gull populations classified as subspecies of Herring Gull by some authorities and distinct species by others. The American Herring Gull *Larus smithsonianus*, the Caspian Gull *L. cachinnans* and the Yellow-legged Gull *L. michahellis* are among the forms sometimes regarded as subspecies of Herring Gull, but more often as separate species. The forms *L. argentatus argentatus*, found in Britain, and the more easterly *L. a. argenteus*, are agreed by all to be Herring Gulls, and across Europe their population together is estimated to be in the region of 790,000 pairs.

Cliff-nesting Herring Gulls are not good neighbours to other seabirds like Guillemots, as they may take eggs and chicks.

The 'natural' breeding grounds of this species are coastal moorland, clifftops, sand dunes and islands on coastal lagoons – any sites that offer some protection against predators or where predators are naturally uncommon. In seaside towns large colonies live on rooftops, wherever ledges and chimneys offer enough level space for them to construct their nests. They forage on beaches of all kinds and out at sea, as well as on town streets, and in winter feed inland at refuse tips, and on fields, lakes, parkland and other open lowland habitats.

ASBO GULLS

Peruse the letters page of a seaside town's newspaper for long enough and you are sure to find a letter calling for a wholescale cull of the local Herring Gull population. The conflicts between gulls and people can be considerable – the birds make a mess with their rooftop nests, sometimes show aggression towards people and pets that they consider to be too near their nests, can spoil a day out at the seaside with their relentless attempts to steal food, and spread rubbish around by tearing open bin bags. Until 2010 Herring Gulls were on the general licence in England and could be humanely killed, and their nests could be destroyed, if they were causing a persistent problem that threatened public health. Their declining population has led to the restriction of the licence conditions, and now those wishing to control adult gulls must apply for an individual licence; however, it is still permitted for landowners to destroy Herring Gull nests under a general licence.

The problem lies in the fact that our seaside towns offer gulls a very acceptable alternative to more rural coastal living. Rooftops provide secure nest-sites, and deliberately feeding the birds takes away their wariness of people and makes them more inclined to make a nuisance of themselves. We unintentionally provide them with another rich food source in the form of edible litter.

Culling might offer a short-term solution, but as long as towns provide productive breeding habitat, more gulls will move in to replace those that are killed unless a cull is extremely severe and wide ranging. Prevention is, as always, better than the cure. It is possible to gull-proof rooftops, and more discipline with what we throw away and how we do it would reduce the gulls' food supply. However, many seaside dwellers love the gulls and welcome their presence, and even most of their detractors would not want to see them removed entirely. With the species in real trouble in parts of its British and Irish range, perhaps we all need to be more tolerant of urban gull colonies.

BEHAVIOUR AND DIET

As is typical in gulls, this is a gregarious bird at all times, although breeding pairs do defend a small territory. It is very common to see groups of Herring Gulls loafing on the beach or close inshore on the sea, or bathing in coastal pools. They also flock with other gull species and forage together, although there is often much squabbling over food items, both on the ground and in the air. Inland Herring Gulls commute between reservoirs and lakes, where they roost in the mornings and evenings, travelling in V formation flocks. In the winter they may arrive at their roosts well before dusk, and spend time bathing and preening before they settle for the night. They are vocal at all times, but especially when on territory.

Herring Gulls catch live fish and other small marine prey by diving from

This gull relishes fish, but is probably more likely to squabble over scraps thrown from fishing boats than catch its own.

Dropping a mussel is a quick way to break its hard shell, a fact which intelligent gulls frequently exploit.

flight or dipping while swimming. However, they also forage along tidelines, in rock pools and on fields, where they attract earthworms to the surface by 'dancing' on the ground to simulate the sound of rainfall (see also page 142). In seaside towns they find rich pickings anywhere litter may be dropped. In some towns they cause problems by tearing open bin bags put out for collection, and may visit bird tables if they can gain access to them. Some individuals are so bold that they steal food from diners at outdoor restaurant tables, or snatch it directly from people's hands. One enterprising if lawless Herring Gull was famously filmed walking into a shop and stealing packets of crisps, which it carried out on to the pavement to rip open and enjoy the contents.

The diet of the Herring Gull is very diverse. Live fish, crustaceans and molluscs are readily taken, with the latter often being dropped onto rocks from a height to break their shells. Herring Gulls sometimes catch and kill other birds and small mammals, especially those that are already injured, will raid other birds' nests for their eggs and chicks, and scavenge carrion of every kind. In some areas they are significant predators of ducklings and wader chicks. They follow fishing boats at sea for discards and scraps, and loiter around the boats when they come ashore for the same reason. They eat invertebrates, including winged ants that they catch in flight, and rifle through rubbish of all kinds in search of anything edible.

BREEDING

Herring Gulls usually pair for life and stick to the same nest-site if possible, whether this is on the ground or in an elevated position. They begin to breed in their fourth year or later. The size of their territory may be just 5m square, but rooftop nests are usually more spaced out than this for architectural reasons. They occupy nest-sites from midwinter and can be seen flying to the nest carrying large quantities of nesting material, including grass clumps, seaweed and sometimes man-made materials like bits of fishing net. Other Herring Gulls that attempt to land at the nest-site are vigorously repelled by one or both of the resident pair.

Vocalisations commonly heard in the breeding season include the 'long call', performed by bowing the head deeply, then throwing it back while giving a series of loud 'kyow' notes. This call is also sometimes given in flight, and serves to advertise territorial ownership. Pairs 'converse' together with low cackling notes, and copulation is often preceded by the pair standing together and taking turns to give a thin 'weeeeoh' call while tossing their heads. The male also feeds his mate as part of courtship, after she approaches him with a soft begging call in the manner of a hungry nestling.

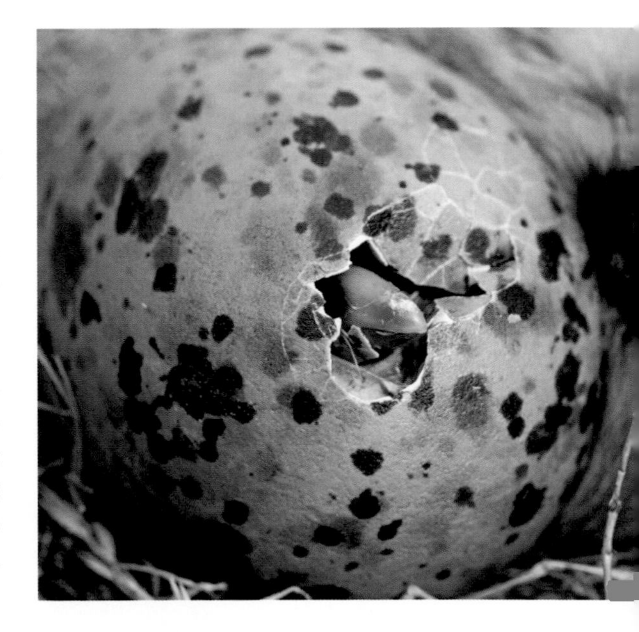

Both the eggshell and the chick which will emerge have a brown, speckled pattern for camouflage.

SUPER-STIMULUS

The Dutch biologist Niko Tinbergen was the one who discovered that Herring Gull chicks instinctively peck at the red spot on a parent's yellow bill to stimulate the parent to regurgitate food. He tested his hypothesis by offering gull chicks a selection of painted sticks, with different background colours and spot colours. Many other large gulls, including Lesser and Great Black-backed Gulls, have a similar bill pattern and show similar behaviour.

Tinbergen also carried out work on what stimulates a Herring Gull to incubate its egg. He made a series of dummy eggs from plaster, which varied in size, colour and amount of spotting, and placed these near nesting Herring Gulls. He found that larger, more richly coloured and spottier eggs were more attractive to incubating gulls. If they were presented with a giant plaster egg with strong colours and many dark spots, they would push their own eggs aside in their attempts to incubate the supersized fake egg. The discovery of a strong response to exaggerated stimuli has since been found in many other animal species.

The red spot on a Herring Gull's bill.

The clutch usually consists of three eggs but sometimes four or two, and both members of the pair incubate the eggs for 28–30 days. Both the eggs and chicks are well camouflaged, with dark mottling on a lighter brown background. Chicks are mobile soon after hatching, and if space allows move away from the nest to find separate places to hide, among long vegetation where they are concealed from passing predators.

The downy chicks grow rapidly and are left unattended for long spells when a few days old. On the rural coastline in some areas they are vulnerable to predation from other gulls and also skuas. In towns a greater danger is posed by overheating if they cannot escape direct sunlight; they may also fall from rooftops and injure or kill themselves. If a chick falls and is unharmed – as is increasingly likely with age because older chicks can slow a fall with their wings – the parents may still continue to feed it on the ground.

Rooftops offer secure nesting sites, and legally cannot be removed unless there is a serious risk to health or property.

The fledglings can fly at 35–40 days, and leave the nest-site soon afterwards, often teaming up with other youngsters on beaches where they learn foraging skills from each other. The adults may also leave for the winter, but in some areas linger near the nesting site all year round.

Herring Gulls would find towns much less suitable habitat if people were careful not to feed them or to discard edible waste.

MOVEMENTS AND MIGRATION

Many of the Herring Gulls that breed in Britain move only short distances, but some travel inland to exploit feeding opportunities, and they are joined by immigrants from further north and east. Ringed birds from Denmark, Germany, Poland, Norway, Finland and western Russia have all been found in Britain in the winter, along with some from France, Spain and Italy.

Young birds from Britain and Ireland may disperse some distance in their early years. Interesting ringing recoveries include a bird found in the Faroes, another in Greenland, and others out at sea in the Bay of Biscay and north Atlantic. After a period of exploration, however, the birds often eventually return to breed at the site where they were born. The oldest British-ringed bird was hatched at Rockcliffe Marsh in June 1978, and found dead at that same site in June 2009, just 16 days shy of its 31st birthday.

THE FUTURE

This gull is still numerous in Britain and there are many places where it is such an obvious and constant presence that it is difficult to believe the species is in real trouble. However, surveys show a clear pattern of decline, with a fall of 48 per cent in 1970–1988, a drop of 13 per cent in 1988–2002, and an apparent further decline (although not based on full national surveys) of 36 per cent in 2000–2012.

A contributory cause of the decline is thought to be botulism, a disease caused by the bacterium *Clostridium botulinum*. This organism thrives on the refuse tips on which many Herring Gulls feed, and also in muddy wetlands. Botulism is usually fatal and is as likely to affect adults as pre-breeding youngsters. Reduced food supplies are also a probable factor behind the decline. Fisheries are less productive and thus generate less waste for the gulls to eat. In addition, landfill site management has undergone various changes over recent years to improve efficiency, and these measures are likely to be reducing the amount of food available to scavenging gulls. On breeding grounds in western Scotland, predation of nests by American Mink is thought to be behind some of the declines of Herring Gulls and other ground-nesting seabirds.

Gulls in urban colonies are generally more productive and survive better than their more rural cousins. Should conditions improve for Herring Gulls in the wider countryside, town colonies could become a source of new recruits for recovering sea-cliff colonies.

Away from towns, the Herring Gull is declining rapidly as a British breeding bird.

Yellow-legged Gull *Larus michahellis*

This gull is very similar to the Herring Gull, and is still sometimes classified as a Herring Gull subspecies (and known as the Yellow-legged Herring Gull). It has a more southerly distribution than the Herring Gull, breeding in south-west Europe, but visits the British Isles in winter, as well as breeding in very small numbers in the south.

INTRODUCTION

A large gull, this bird has the sturdy build of a Herring Gull, but has yellow rather than pink legs and a red rather than yellow eye-ring; the grey parts of its plumage are a few shades darker, intermediate between those of the Herring Gull and Lesser Black-backed Gull. In winter it has less brown head streaking than either of these species. Identifying it in juvenile and first-winter plumage is rather difficult, but it usually looks paler-headed than either the Herring Gull or the Lesser Black-backed Gull.

DISTRIBUTION, POPULATION AND HABITAT

This is a very rare breeding bird in Britain, with only one or two pairs breeding each year, plus a handful of mixed pairs comprising a Yellow-legged Gull paired with either a Herring Gull or a Lesser Black-backed Gull. Mixed pairs often breed successfully and produce hybrid young. Most records come from the Dorset coast, where the species first bred in 1995. In winter visiting Yellow-legged Gulls from the near Continent arrive in increasing numbers, with a little over 1,000 birds each year on average. These are distributed widely across England and Wales, both inland and on the coast.

The darker upperside and yellow legs help distinguish this species from the very similar Herring Gull.

The taxonomy of this gull and its near relatives is still unresolved, although it is usually considered to have two subspecies, the nominate breeding around the Mediterranean (including parts of the Middle East and North Africa), and *atlantis* occurring along Atlantic coasts from Portugal to Morocco. The Armenian Gull *L. armenicus* of the Middle East is also sometimes grouped with the Yellow-legged Gull, as is the Caspian Gull *L. cacchinans* and its associated subspecies. With all of this confusion and disagreement, no useful estimate has yet been made of the Yellow-legged Gull's population size.

The species breeds mainly on the coast, using shingle islands in lagoons and also rooftops. Wintering birds are seen at refuse tips and with other gull flocks on reservoirs.

BEHAVIOUR AND DIET

The Yellow-legged Gull's habits are similar to those of the other large gulls, although it seems a little more aloof by nature and more likely to be seen alone. It is more often seen among Lesser Black-backed Gulls than Herring Gulls. Like the other large *Larus* species it is a generalist feeder, catching various kinds of live prey but also scavenging for carrion, especially on the tideline, and scraps at rubbish dumps.

BREEDING

The species' breeding behaviour is very like that of Herring and Lesser Black-backed Gulls. It has its own slightly different version of the territorial 'long call', and there are other subtle differences in vocalisations and other interactions between a pair. However, these are not enough to prevent it from successfully forming pair bonds with those other two species, although if there were more Yellow-legged Gulls present in summer there would probably be fewer or no mixed pairs.

The female lays two to four eggs in a large nest made from assorted vegetation, and she and her mate share the 28–30-day incubation period. The chicks take 35–40 days to fledge, but leave the nest at about three days old to hide nearby (if there is enough space for them to do so).

MOVEMENTS AND MIGRATION

Most of the Yellow-legged Gulls that visit Britain and Ireland in winter are young birds, not yet old enough to breed, but there are a few adults, too. There are several colour-ringing schemes in Britain and abroad that focus on gulls, and birdwatchers with a special interest in gulls are very diligent at checking ring numbers. Data from these sources should help in achieving an understanding of the movements of this species. For example, the North Thames Gull Group has in 2012 and 2013 found two Yellow-legged Gulls that were originally ringed in Germany, and has ringed more than 60 individuals in Essex since 2005 – a good tally considering that this is still a rare bird in Britain.

This is a rare but regular visitor to all kinds of gull-attracting habitats in southern Britain, from landfills to urban park lakes.

THE FUTURE

Although this species has been breeding in Britain since 1995, numbers have not increased in this time. Whether they will do so in future is difficult to predict, but the Yellow-legged Gull is increasing and spreading on the Continent so there is a supply of 'new' birds that will need to find places to breed. With wintering numbers increasing, more young birds are gaining the opportunity to assess sites in England and Wales for breeding potential.

The Yellow-legged Gull has in the past been culled in parts of its range to reduce predation on other vulnerable species (Audouin's Gull in Spain, and the White-faced Storm Petrel on islands off Iberia). It is a protected species in Britain, but as it is a new colonist with a very small pioneer breeding population, no special measures are in place to conserve it.

Great Black-backed Gull *Larus marinus*

Huge and powerful, with a character that in no way belies its formidable appearance, the Great Black-backed Gull is an impressive and imposing bird with a rather stately bearing. It is the largest gull species in the world and probably the most predatory. It is usually present in only small numbers in any mixed gathering of gulls, and immediately draws attention because of its size and the adults' very black wings. Like our other large gull species it is undergoing a population decline, which has picked up speed since the year 2000, and it has been moved into the Amber category of conservation concern in the UK.

INTRODUCTION

The sheer size and bulk of this bird helps to separate it from other gulls in all plumages, especially in mixed flocks. It is clearly larger than a Herring Gull and has a sturdy, powerful build with proportionately short wings, giving it a more short-bodied look when perched, compared with the much longer winged Lesser Black-backed Gull. The bill is also strikingly large and heavy. Adults have almost jet-black backs and wings that show no contrast with the black wingtips. The outermost primary has a large, solid white tip. The legs are pink, brighter in breeding plumage, and in winter the bird develops only very light brown head streaking.

Juveniles and first winters are best identified by size and shape, and are clearly paler headed than juvenile Herring or Lesser Black-backed Gulls, although they can be confused with young Yellow-legged Gulls. In flight the Great Black-backed Gull looks heavy and barrel-chested, and its wingbeats appear more laboured than those of smaller species. It also has a deeper voice than those of the other large gull species.

The massive bill, proportionately small eye, and pure white tip to the outermost primary feather help identify this gull.

AN ENDURING NAME

The Swedish botanist Carl Linnaeus, who lived from 1707 to 1778, was one of the first people to make a concerted attempt to devise a consistent and comprehensive classification system for all living things. He also devised the binomial scientific naming system that we use today, whereby each species has a genus name (such as *Larus*), which is shared by all closely related species, followed by a species name (such as *marinus*) which is unique to a particular species. Linnaeus assigned a binomial scientific name to all the animal and plant species that were known to him.

Over the years since Linnaeus's time, we have constantly built on our knowledge of how species are interrelated and this process is ongoing today, with modern techniques allowing us to explore relatedness at the genetic level. In the course of these investigations we have discovered that many species need to be moved from one genus to another, and determined that some genera need to be split into two or more. Organisms' scientific names have been changed accordingly to accommodate this – some species have gone through many different scientific names as our understanding of their relationships with other species has developed. Also, many more species are now known to science than were in the 18th century – in his 1754 master work *Systema Naturae* Linnaeus classified 564 species of bird, and there are now known to be about 10,000. However, a handful of species still retain their original Linnaean scientific names, one of which is the Great Black-backed Gull. The meaning of its name has a pleasing simplicity. Birdwatchers are fond of explaining to others that there is no such species as a 'sea gull' but if meanings of scientific names are taken into account then there is, because *Larus marinus* means exactly that.

DISTRIBUTION, POPULATION AND HABITAT

Great Black-backed Gulls breed around much of the coastline of Britain and Ireland, the largest gaps being in eastern England and south-east Ireland. The highest numbers are on Orkney, Shetland and other Scottish islands, on northern and western mainland Scotland, and in western Ireland, the south-west and west Wales. They are not as colonial as other gulls and many pairs breed alone; colonies also tend to be smaller than those of other gulls, with only a few holding more than 500 pairs and many fewer than 100. There are a little under 20,000 pairs in Britain and Ireland, about 17,000 of these being in the UK. In winter numbers increase to about 77,000 individuals, which are mainly on the coast but also inland in parts of England, Scotland and Wales.

Elsewhere in the world this gull breeds in Iceland, north-west France, Scandinavia, the Baltic states and north-west Russia, and also on the Atlantic coast of the USA and Canada, and in southern Greenland. In winter it spreads south as far as Portugal in Europe and the Caribbean in the Americas. The world population is in the region of 170,000 individuals, so the British Isles hold quite a significant proportion – just under 10 per cent.

Great Black-backed Gulls are strictly seaside birds when it comes to choice of nest-sites.

This gull is more strictly coastal than the other common large gull species in Britain, and nests mainly on undisturbed rocky coastlines and on moorland close to the sea. A few pairs breed on rooftops, although the species has not adopted this habit to anything like the same extent as the Herring Gull or even the Lesser Black-backed Gull. A small number of birds also nest inland, primarily on shingle islands in lakes, large rivers and reservoirs. In winter the species ranges more widely and visits other coast types, including open shingle and sandy beaches, and seaside towns (non-breeding birds including youngsters that are not of breeding age may occur at such places all the year round). It is less common inland than other large gulls, but a few can often be found in major roosts and at refuse tips.

A flock of mixed-age Great Black-backed Gulls commutes to their roost on a winter afternoon.

BEHAVIOUR AND DIET

The Great Black-backed Gull is not as strongly gregarious as other gulls, and it is not unusual for it to nest, feed, rest and travel alone, or just with its mate. In mixed-species flocks it is nearly always heavily outnumbered by other species. It has a subtly slower and more deliberate way of moving that (along with its size) makes it stand out in an active flock, although it can move at high speed when necessary. Very few other animals pose a threat to this gull – only large eagle species stand a chance of killing it. In any mixed feeding situation it easily dominates most other species when competing for food, at a washed-up carcass for example. Among British seabirds only the Great Skua may be able to stand up to it. Its flapping flight is ponderous and its take-off is more protracted than that of smaller species, but it has great endurance on the wing and also soars on thermals to gain height.

The species has a wide array of foraging methods. It splash dives and surface dips for fish and other swimming organisms, and chases injured birds and mammals on foot. It turns over tideline debris to look for washed-up dead fish and other carrion, and pursues fishing boats to snatch thrown-away scraps. Mussels and other molluscs picked from rock pools may be dropped from a height to break them, although smaller molluscs can be swallowed whole. It can catch less agile birds in flight, often then carrying them to the sea to kill and consume them, and also takes eggs and chicks from other birds' nests. It chases other gulls when they are carrying food, to snatch their meals out of their bills in mid-air.

Washed-up dead fish and other sea animals provide a major part of this species' very varied diet.

The diet of this bird is, as might be expected, very diverse, including living prey and carrion of all kinds, and scavenged scraps. In the breeding season some individuals concentrate on one or just a few prey types; for example, on Dun, St Kilda, several pairs are specialist Puffin hunters, catching these birds in flight or loitering near nest-sites to catch them as they return to their burrows. On Skokholm island some of the breeding Great Black-backed Gulls hunt Manx Shearwaters by preference, while others take mostly Rabbits. Less often the gull takes even larger prey, and it has been known to catch and kill other large gulls and Jackdaws. Most birds are too quick for it, but newly fledged youngsters and any birds that are injured are quickly caught and killed. The species can account for significant mortality of both adult and young birds of various species at some seabird colonies.

Performing its long-call, a Great Black-backed Gull shows off its huge gape, wide enough to swallow a petrel or other small seabird whole.

BREEDING

This gull breeds either singly or in relatively small colonies, and like related species, shows a high degree of fidelity to both its nest-site and its partner. The type of rugged coastline it prefers for nesting may have large spaces between suitable level spots where a pair can build its sizeable nest, and each pair needs a territory that covers several square metres. The males are first to occupy territories, during the winter, and they call to attract a female – with luck their mate from the previous year but if she fails to return then a different female. The advertising 'long-call' is given either from the ground or in flight.

Within the territory the paired-up adults make several initial scrapes before selecting one to develop into the nest for that year. Ideal sites offer good visibility but some shelter from the prevailing wind. However, rooftop nesters re-use nests year on year, although some repair work may be necessary. Both members of the pair collect nesting material, ferrying bill-loads of torn-up muddy grass, seaweed and other vegetation to the chosen spot, where they shape it into a nest about half a metre across. As nest building nears completion, courtship feeding from the male to the female becomes frequent.

A typical Great Black-backed Gull clifftop nesting site, with one adult on eggs and a brood of half-grown chicks nearby.

FOUR-YEAR PLAN

Gulls take two, three or four years to transition from juvenile to adult plumage. Large species like the Great Black-backed Gull are 'four-year gulls'. The names usually used for the different plumage stages are juvenile, first-winter, second-winter and third-winter, although in reality the annual moults (a full moult in autumn and a partial moult, of body feathers only, in spring) are quite protracted, so one 'first-winter' may look quite different from another, and bleaching of older feathers from exposure to the sun can also change a bird's appearance quite dramatically.

The juvenile plumage is very short-lived, being moulted through the bird's first autumn to the slightly paler, but still entirely brown, first-winter plumage. In general it is the back or mantle of the bird that begins to acquire adult colour first, in the second winter, and at this time the eyes have also changed from dark brown to pale. At the same time the head and body become much whiter, although they are still extensively streaked. By the third winter the gull is quite adult-like but retains juvenile-like brown mottling on the wings, and a dark band to the tail. The bill is mostly yellow by this point, but with some black near the tip. Large gulls in their fourth winter are virtually indistinguishable from adults, but in the case of Great Black-backed Gulls it is still another two years or so before they are ready to breed.

A first-winter Great Black-backed Gull. At all ages, the huge bill is a good clue to identification.

The clutch consists of two or three greenish, mottled eggs. Laying is spaced out over four days or longer, and incubation begins immediately, so that the chicks hatch asynchronously, after about 27 days of incubation each. The hatchlings are appealing balls of grey-brown fluff, with open eyes and strong legs, and very soon they leave the nest and spread out separately within the territory, hiding in cover except when being fed by a parent. The adults may brood them at night and in cold conditions.

A breeding pair of Great Black-backed Gulls makes a formidable team when it comes to repelling predators. However, because the birds often nest away from other pairs and indeed other birds, they do not have the same 'back-up' as birds in large colonies, and nests may be left unattended at times. Eggs and chicks may then fall victim to opportunistic predators such as skuas, crows and rats. However, the young birds grow quickly and soon they are a force to be reckoned with. They can fly at about 50 days old, and soon after the first flight are fully independent.

A Great Black-backed Gull will usually win a fight over food with another gull, unless that other gull is also a Great Black-backed Gull.

MOVEMENTS AND MIGRATION

Many British and Irish breeding Great Black-backed Gulls are residents and spend the winter close to their nesting grounds. They are joined by visitors from further north or east, along the coasts of France, Belgium, the Netherlands, Denmark and Norway in particular. Juveniles from Britain and Ireland usually move south and winter on the Atlantic coasts of France and, in smaller numbers, Spain, where they may remain until they reach breeding age.

THE FUTURE

At the end of the 19th century the Great Black-backed Gull was a very rare bird in Britain and Ireland, because it had suffered heavy persecution; there may also have been other causes of decline, which are not known. A count in England and Wales in 1893 produced just 20 pairs, but this had grown to 1,000–1,200 by 1930. Its numbers continued to rise through most of the 20th century, but there was a small decline of 7 per cent between the Operation Seafarer survey in 1969–1970 and the subsequent Seabird Colony Register in 1985–1988. By Seabird 2000 the population had fallen by another 4 per cent overall, although this hides a very mixed picture of change whereby many good-sized colonies entirely disappeared over that interval, while several small new colonies grew rapidly. Data from various colonies, collected for the Annual Seabird Monitoring Programme, indicates that the overall rate of decline has accelerated considerably since the year 2000.

It has been suggested that the Great Black-backed Gull is now suffering from the same problems of reduced food availability, especially fishery discards, as Herring and Lesser Black-backed Gulls, but has taken longer to be affected because it enjoys a competitive advantage over the other two species. As a predator on other seabirds, it may be declining (at least in some areas) in response to a general decline in seabird numbers. In Wales quite high numbers were affected by botulism in the late 20th century.

In response to its decline, the Great Black-backed Gull can no longer, as of 2010, be legally killed or have its nests destroyed under a general licence – those with a justifiable need to control it must apply for an individual licence. This is in response to the species' recent decline, but is unlikely to make much difference to the species' population. Studies at breeding colonies suggest that adult survival has not changed significantly but that breeding productivity has fallen, so the problem seems to lie with insufficient food supplies during the breeding season rather than in winter.

Flying past a seabird cliff, a Great Black-backed Gull is always looking out for the chance to grab an unattended chick.

Other gulls

Not only does the world's largest gull occur in Britain, but the smallest one does as well, in the form of the **Little Gull *Hydrocoloeus minimus***. This charming little bird, about the size of a Blackbird, does not breed here, although the occasional pair has turned up within a Black-headed Gull colony and attempted to nest. It occurs primarily as a passage migrant, and can be quite numerous in spring and autumn at certain favoured sites. It may also be seen in smaller numbers at other times of year. It breeds in north-east Scandinavia and further east, with just a few isolated colonies further west, and there are also a few on eastern North American coasts following the first breeding there in the early 20th century. The European birds overwinter on Mediterranean and Atlantic coasts.

This gull has a markedly dainty appearance, especially when seen alongside other gulls, and has a very light, tern-like flight. It feeds in the manner of a tern as well, dipping to the water's surface to pick insects from near the surface. In summer plumage it shows a full blackish hood without any white markings around the eyes, and the wing feathers are tipped white, giving the entire wing a narrow white trailing edge. The upperside of the wing is light silver-grey, but the underwing is strikingly dark and dusky, and there is sometimes a faint rosy flush on the breast and belly. The bill looks very small, even accounting for the size of the bird. In winter the hood mostly disappears. The bird then has a mostly white head, sometimes with a dark patch on the rear crown, and has a dark spot behind the eye. Juveniles and first-winters have a grey-and-black pattern that is quite similar to a juvenile Kittiwake's, but without the black collar and with a dark central smudge on the cap.

The smoky, white-edged underwings of the Little Gull aid identification in all plumages.

Iceland Gulls are scarce winter visitors, but can reach double figures at northern fishing harbours.

After breeding, Little Gulls move east, gathering in sheltered bays and coastal freshwater bodies for their annual moult. At this time there may be gatherings of dozens, occasionally hundreds, at east coasts. They then mainly move further west, with large numbers wintering in the Irish Sea. On return migration they often go overland and can be seen in the Merseyside area as they head towards Finland, the Baltic states and Russia. Most foreign-ringed birds recovered in Britain have been from Finland.

Numbers of breeding Little Gulls are increasing quite strongly in the colonies closest to us in Scandinavia. This trend may fuel a westwards push and bring about more breeding attempts in Britain in the future.

Only the newest British bird field guides make mention of the **Caspian Gull *Larus cachinnans***. This bird was until 2007 classified by the British Ornithologists' Union as an eastern form of the Herring Gull, while some authorities split it from the Herring Gull but 'lumped' it with the Yellow-legged Gull. As genetic research has made headway with the unpicking of the complicated interrelationships of the 'large white-headed gull' group, so interest among birdwatchers has grown. It has become clear from the patient efforts of the keenest gull-watchers that Caspian Gulls regularly visit Britain in small numbers.

This gull is very like a Herring Gull in general appearance. It has a subtly different general look, with a slimmer shape, longer legs and a long-looking face with a long, parallel-edged bill, less angular-looking than a Herring Gull's. Its rear end is held in a rather drooping stance, and its eyes look strikingly small – they are also darker than Herring Gulls' eyes. Its grey parts are intermediate between the paler grey of the Herring Gull and the darker tone of the Yellow-legged Gull, and its legs are a rather dull flesh tone. In immature plumages identification is even more of a challenge. Caspians usually look paler headed than similar-aged Herring and Lesser Black-backed Gulls, and the small eye and 'long' head may also be helpful features to note. However, there is individual variation and some birds will never be safely identifiable.

Caspian Gulls breed around the Caspian Sea and Black Sea, and eastwards across Asia as far as north-west China. They are spreading westwards and have established breeding populations in Poland

and Germany; this expansion, as well as improved observer awareness, could explain why sightings in Britain are on the increase. The species is mostly seen in south-east and eastern England and the Midlands, and is best looked for among large gull roosts and gatherings in rubbish dumps. At such places there should be plenty of Herring and Lesser Black-backed Gulls, and with luck a few Yellow-legged Gulls as well, allowing for side-by-side comparison.

The term 'white-winged gulls' covers two species of the High Arctic that visit us in winter, and look similar to Herring Gulls but have pure white rather than black wingtips. The smaller of the two is the **Iceland Gull *Larus glaucoides***, which in contradiction to its name does not breed in Iceland (although it does winter there). Its breeding range extends across Arctic Canada to Greenland.

This is a slim, long-winged gull, a little smaller than the Herring Gull but with a much less bulky look and a rounder, gentle-looking head. Its overall shape is more reminiscent of a Lesser Black-backed Gull or even a Common Gull, with wingtips reaching well beyond the tail-tip. In adult plumage it is white apart from the very light grey wings and back, and it is also very light coloured in immature plumages. First-winter birds have off-white plumage with mid-brown spots and streaks all over, so from a distance the plumage looks the colour of very milky tea. Second-winters are almost pure white, especially towards spring when their worn plumage has become quite bleached due to exposure to sunlight, and third-winters are very similar to adults. All birds show almost unmarked white wingtips at all ages.

Iceland Gulls move south from their breeding grounds in winter, and may show up anywhere in the British Isles, although they are most numerous further north. In some winters there are large influxes, and productive feeding grounds such as fishing harbours in northern Scotland and eastern Ireland can attract double-figure numbers. They also sometimes turn up inland, in mixed gull roosts. Total numbers in the British Isles each winter are in the region of 240 birds, a figure that has risen since the year 2000.

A different form of Iceland Gull that is sometimes found in Britain is **Kumlien's Gull *Larus glaucoides kumlieni***. This bird is a little darker than the Iceland Gull, with some mid-grey areas in the wingtips, but otherwise it is very similar. Its taxonomic status has been the subject of much debate, with it variously being considered a subspecies of Iceland Gull, a full species in its own right, or a stable population of hybrids between the Iceland Gull and the darker winged **Thayer's Gull *Larus thayeri*** of North America. The preferred theory at the time of writing is that it is an Iceland subspecies, but it is a notably variable gull, which does support the hybrid theory.

Kumlien's Gull is a vagrant from North America, and has a rather uncertain taxonomic position.

The largest 'white-winged' gull, the Glaucous Gull approaches the size and bulk of a Great Black-backed Gull.

The Iceland Gull's big brother is the **Glaucous Gull** ***Larus hyperboreus***. This gull is almost identical to the Iceland Gull in plumage at all ages, and the two species can be confused. The average Glaucous Gull is, however, considerably larger than the average Iceland Gull and indeed larger than the Herring Gull, with the largest birds approaching the size of Great Black-backed Gulls. The Glaucous Gull is also a differently shaped bird, with a broad, stocky frame and proportionately much shorter wings, barely projecting beyond the tail. Its head is more angular and its expression more severe than the Iceland Gull's. In first-winter plumage it looks a little greyer and 'grubbier' than the warm biscuit-coloured Iceland. First- and second-winters have a distinctive bill pattern, with a pink base and very neatly marked black tip, which can be helpful when trying to pick out a young Glaucous in a large flock of other subadult gulls.

The Glaucous Gull has a more extensive range than the Iceland Gull, breeding across Arctic Eurasia as well as North America. However, in most years it is slightly less numerous in the British Isles than the Iceland Gull, with 170 records each winter on average, although this is a recent change, with Glaucous records declining as Iceland Gull records increased over the late 20th and early 21st centuries. Its distribution pattern is almost identical to that of the Iceland Gull, with more records further north. Often the two species can be seen together at favoured sites, such as harbours in Shetland and Orkney. Individual Glaucous Gulls show site fidelity to their wintering grounds, so may be seen year after year at the same spot.

Although most Glaucous Gulls are found on the coastline, they will also go inland and can be found foraging at rubbish tips alongside other gulls. Being larger than the other species (excepting the Great Black-backed Gull, which is scarce inland), they can easily dominate proceedings and commandeer the best food items from the other birds around. As well as scavenging they sometimes catch living prey, including fish but also occasionally other birds.

Even in brown first-winter plumage, the Glaucous Gull still shows obvious whitish wingtips.

Glaucous Gulls sometimes hybridise with Herring Gulls, producing offspring that are intermediate in appearance between the two species, which usually show some distinct brown or black in the primary tips. Any large gull that is absolutely pure white is likely to be a Herring Gull or other common species with leucism (a plumage abnormality featuring a complete absence of melanin pigment in the feathers); even the most worn and white second-winter Glaucous or Iceland Gull still shows a suggestion of darker speckling.

Birders can and do spend their life savings to travel to remote northern beaches when an Ivory Gull turns up.

There is a naturally pure white gull that occasionally visits Britain. The **Ivory Gull *Pagophila eburnea*** is a High Arctic species found in Eurasia, North America and Greenland. It is a vagrant to mainly the far north of the British Isles. It is a smallish and compact gull, with entirely white plumage when adult. Immatures are white with some black spotting on the wings and around the face. Because of its beauty and rarity, it is much sought after by birdwatchers. Many of the Ivory Gulls found in Britain and Ireland were located when they were attracted to a whale or seal carcass on a remote shoreline. The species is declining, to the extent that it is considered Near Threatened, so its appearances on our coasts may become even more sporadic.

Another beautiful Arctic gull that is a rare vagrant to the British Isles is **Ross's Gull *Rhodostethia rosea***. A small and dainty gull with an unusual diamond-shaped tail, the adult of this species has white plumage with light grey wings, and a complete narrow black collar. It often has a distinct pinkish flush to its head and underside. In winter the collar is lost, and it then looks rather like a winter-plumaged Little Gull with its very small dark bill and dark eye-spot. The first-winter bird has a Kittiwake-like black zigzag pattern across its wings. Ross's Gull breeds in Arctic North America and Siberia.

Strong gales in mid-autumn can force various interesting seabirds to our shores and inland, including a classic trio of species that are rarely seen in Britain and Ireland except in the wake of storms. These are the Grey Phalarope (see page 106), Leach's Petrel (see page 71) and **Sabine's Gull *Xema sabini***. Looking rather like a Black-headed Gull in breeding plumage, Sabine's Gull is an unusual and very pelagic gull with a clearly forked tail, which breeds in the Arctic and migrates to Africa or South America in winter. Adults in breeding plumage have a solid sooty grey-black hood, which becomes just a dark collar in winter; the rest of the body is white with dark grey wings. The bill is dark with a distinct yellow tip. In flight the wings show black outer primaries, white inner primaries and secondaries, and grey coverts. These blocks of colour form three contrastingly toned triangles on each wing, creating a pattern unlike that of any other gull. First-winters show a similar pattern, although their uppersides are scaled brownish-grey rather than plain grey; they also have a smoky-grey neck 'boa'.

With its little bill, long wings and rose-flushed chest, Ross's Gull is perhaps the prettiest of all gull species.

Autumn gales in the south-west bring the chance to see the uniquely patterned Sabine's Gull, usually in first-winter plumage like this one.

Between 100 and 200 Sabine's Gulls are recorded in Britain each year, most of them by seawatches as they go by offshore, and the pattern of distribution has a southerly and westerly bias. After very severe weather a few Sabine's Gulls are usually discovered inland, at reservoirs or in the lower reaches of rivers, where they can be watched gliding and dipping at the water in the manner of terns, feeding to restore their strength before resuming their migration down the Atlantic.

Two medium-sized species of gull that breed well south of the British Isles have occurred a handful of times in Britain. They are the **Slender-billed Gull *Larus genei*** and **Audouin's Gull *Ichthyaetus audouinii***. These species are superficially similar, with white bodies, black-tipped grey wings and red bills, but are not especially closely related. The Slender-billed Gull has an extensive but patchy distribution across north Africa, west and central Asia and the eastern Iberian Peninsula. It is a slim, leggy gull, which looks somewhat like a large, long-billed and pale-eyed winter-plumage Black-headed Gull. Audouin's Gull has the distinction of being one of the world's rarest gull species, although its numbers are recovering strongly from a low point of just 1,000 pairs in the 1960s. Conservation

Once a globally threatened species, Audouin's Gull is increasing in its southern European breeding grounds but it is still a rare visitor to Britain.

measures have proved highly effective and its numbers have grown to about 22,000 pairs, distributed all around the Mediterranean. It is a medium-sized gull with frosty grey wings and a quite heavy red bill, marked with a black ring. It has occurred in Britain a handful of times, all since the year 2000.

The neat black bill-ring of the Ring-billed Gull is crisper than the similar marking on winter-plumaged Common Gulls.

The **Great Black-headed Gull *Larus ichthyaetus***, or Pallas's Gull, is a very impressive bird, the largest 'hooded' gull by far. It breeds in a few scattered localities in eastern Europe and Asia, the closest birds to us being in Ukraine, and migrates further south for winter. Only a little smaller than the Great Black-backed Gull, this bird has a full black hood in breeding plumage, and is otherwise white with grey black-tipped wings and a very sizeable red bill. It has occurred only once in Britain since 1950, and sadly is declining and therefore even less likely to occur again in future unless its fortunes improve.

Another large Asian gull that is about to join the British List is the **Slaty-backed Gull *Larus schistisagus***, a record of which is currently under consideration. The bird was seen at an Essex refuse tip in January 2011, and its identification was testament to the skill and dedication of the keen local 'larophile' who discovered it. This species, which is comparable to a large, dark-backed Herring Gull in appearance, breeds in Siberia, North Korea and Japan, so its occurrence in England is wholly remarkable and perhaps unlikely to be repeated anytime soon.

Many of the gulls that breed in North America have occurred as vagrants in the British Isles. Perhaps the most familiar of them is the **Ring-billed Gull *Larus delawarensis***, a common bird in Canada and the USA. Several are found in Britain and Ireland each winter, many of them returning birds that winter at the same location for several years in succession. The Ring-billed Gull is similar to the Common Gull, but is a little larger with pale eyes and darker grey upperparts, and has a more robust shape with proportionately shorter wings. The black ring on the bill is present all year round, and is much neater and more distinct than the smudgy dark bill-ring that Common Gulls develop in winter. The majority of records are in the west, as would be expected for a bird that arrives here from North America, but some birds are also found on the east coast.

This Ring-billed Gull visited an Essex seaside town for 11 winters, and acquired the nickname 'Rossi' after a local ice-cream parlour.

Largest of several 'hooded' North American gulls that occasionally stray to Britain, the Laughing Gull is named for its chuckling call.

Three species of 'hooded' gull from North America are rare vagrants to Britain. They are, in descending order of size, the **Laughing Gull *Leucophaeus atricilla***, **Franklin's Gull *Leucophaeus pipixcan*** and **Bonaparte's Gull *Chroicocephalus philadelphia***. Bonaparte's Gull, which looks like a smaller, black-billed version of the Black-headed Gull, is probably the most frequent, occurring annually. The Laughing Gull is Common Gull-sized with rather dark slaty-grey wings and a large, slightly droopy-looking bill. Although usually very rare in Britain and Ireland, it has occurred in higher numbers on occasion, for example in autumn 2005 when there was a significant influx, with at least 18 individuals present on one day. Franklin's Gull is the least common of the three. Like the Laughing Gull it is a dark bird, about the size of a Black-headed Gull. It is widespread in North America but breeds inland, which goes some way to explain its rarity as a transatlantic vagrant.

Finally, two large *Larus* gulls from North America have occurred as very rare vagrants in the British Isles. The **American Herring Gull *Larus smithsonianus*** is often regarded as a subspecies of the Herring Gull, and is so very similar to it that identification is always challenging and some birds are probably never noticed. A **Glaucous-winged Gull *Larus glaucescens*** was found in Gloucestershire in 2006, but there have been no further confirmed records. In adult plumage this bird resembles a pale Herring Gull with grey rather than black primary tips – it could be mistaken for a Herring x Glaucous Gull hybrid and is perhaps easier to identify in its rather plain, solid-grey first-winter plumage. It breeds in western North America, from Alaska down to Washington State.

Bonaparte's Gull resembles a slightly undersized Black-headed Gull, and usually associates with that species when it turns up in Britain.

Terns

Terns are renowned for their grace and elegance – the old name of 'sea swallow' aptly describes a tern's slim, fork-tailed outline and aerial skill. They belong to the family Sternidae, part of the large order Charadriiformes, and are closely related to the gulls. Worldwide there are about 44 species in about 12 genera, of which 18 species have been recorded in Britain and/or Ireland. Most terns are true seabirds. However, the marsh terns (genus *Chlidonias*) lead a more inland existence (although they are still found at sea outside the breeding season). Terns are much more specialised than gulls, with most species feeding almost exclusively on fish that they catch themselves, while others catch crustaceans and large numbers of insects.

The typical terns of the genus *Sterna* are very similar in appearance, with a white head and body, grey wings and back, elongated outer tail feathers, a well-defined black cap in breeding plumage, and often a red or yellow bill. Other tern genera are broadly similar, except for the noddies (genus *Anous*, which are dark-plumaged, and *Procelsterna*, which are smoky-grey), and the White Tern *Gygis alba* with its entirely white plumage. Some authorities believe that both the noddies and the White Tern should be classified in separate families.

The terns that breed in Britain nest in large, noisy colonies on flat ground, often on beaches and on islands within coastal lagoons. Inland breeding is rare, except in the case of the Common Tern. For protection in their exposed breeding habitat terns rely on both camouflaged eggs and chicks, and a coordinated defensive response to approaching predators. However, a huge amount of suitable tern habitat has been lost to seaside development, and the colonies that remain can be highly vulnerable to disturbance, predators and the vagaries of the weather. Some of our terns have undergone severe declines in recent years, and the most vulnerable colonies require very proactive conservation management to save them, including measures such as providing anti-predator fencing and shelters for the chicks.

Our terns are all long-distance migrants, heading for Africa for the winter months. On migration they can be seen going by offshore, often flying very near the shore and pausing to splash down into the water in pursuit of fish. Some also may migrate overland, although they are unlikely to make any stop-offs en route, so are rarely seen except by the most dedicated watchers of 'visible migration'.

Arctic Terns in a mid-air argument.

Little Tern
Sternula albifrons

With its small size, distinctive head and bill pattern, and quirky mannerisms, the Little Tern is a charming and usually unmistakable species. Like all our terns it is a summer visitor. It nests in colonies on beaches, and mainly forages inshore close to its breeding site. Its preference for nesting on open beaches alongside calm, shallow sea puts it in significant conflict with human interests, and many important colonies have been lost or greatly reduced because of development and disturbance, leaving it in a precarious position as a British breeding bird. Protecting those colonies that remain is a conservation conundrum that must be solved to safeguard its future.

INTRODUCTION

The Little Tern is our smallest tern species. It looks long-billed with a fairly long, forked tail, but lacks extended tail streamers. Adults in breeding plumage have a black eye-stripe and white forehead-patch, rather than a solid black cap like the other sea terns, and the yellow, black-tipped bill is also unique. The legs and feet are orange-yellow. Little Terns have darker wings than our other breeding terns, with an obvious blackish leading edge from the carpal joint to the wingtip. Adults moulting into winter plumage develop a larger white forehead-patch and duller feet and bill, while juveniles have dark-scaled upperparts and a blackish bill and feet. In flight Little Terns lack the fluid grace of larger terns, instead moving in a very distinctive jerky manner, with rapid wingbeats and abrupt changes in height and direction. The high, chattering and rather thin call, less grating than that of larger terns, is most often given at or near the nest.

A Little Tern in hunting mode, head tilted down to scan the sea below.

As well as its small size, this tern stands out among other sea terns in that it lacks very long tail streamers.

DISTRIBUTION, POPULATION AND HABITAT

The Little Tern is widely but very patchily distributed around the coastline of the British Isles, with important concentrations in East Anglia, the central south coast of England, north-east England, the western Highlands and the Western Isles. Its breeding population in the British Isles is estimated to be about 1,900 pairs, about 60 per cent of which are in England. Those that nest in Britain, which make up less than 3 per cent of the global population, spend winter off the west coast of Africa, probably as far south as South Africa.

Worldwide the Little Tern's range is extensive, with breeding populations across Europe, much of Asia and northern Australasia, and one colony on Hawaii. The southwards migration sees most of the African, South-east Asian and southern Australasian coastlines occupied throughout the austral winter.

The species nests primarily on shingle and sandy beaches in colonies of a dozen or so pairs, occasionally 40 or more, relying on group attacks to drive away predators, and on the camouflage of its eggs and chicks as a second line of defence. It also uses natural islands in coastal saline lagoons. Chosen breeding sites are always close to shallow, warm and quite sheltered sea, where the adults can feed, and it rarely nests in close company with other tern species. Elsewhere in Europe it also nests on shingly river shores and islands, but in Britain it is very rarely seen inland, even on migration.

THE LEAST IDEA

Rye Harbour nature reserve, in East Sussex, has played host to more species of tern than almost anywhere else in the British Isles. It is home (at least in some years) to a breeding colony of Little Terns, and in 1983 this colony drew attention from birdwatchers who noticed a distinctively squeaky voice coming from its midst. When the owner of the voice was tracked down it was initially thought to be just another Little Tern, but detailed examination of its plumage and call eventually revealed that it was in fact a Least Tern *Sternula antillarum*, the North American equivalent of the Little Tern.

Affectionately nicknamed 'Squeaker', the tern visited the colony every summer until 1992. It perhaps did not gain as much attention as it would have done today, as the Least Tern was only split as a separate species from the Little Tern in 2006 – and then only by the American Ornithologists' Union; the British Ornithologists' Union still considers it to be a subspecies of the Little Tern. It is certainly extremely similar to the Little Tern, with a slightly greyer rump and outer tail, but is most easily distinguished by its voice.

Small fish such as sandeels are important in the diet of Little Terns.

BEHAVIOUR AND DIET

A typical tern in its habits, the Little Tern is energetic, fast-flying, tireless and indomitable on the wing. It is most often seen flying close inshore and a few metres up, working its way along parallel to the shore. It then turns on a knife-edge and may retrace its steps, hover or drop straight down to catch prey close to the water's surface. The bird makes a shallow head-first plunge, generally keeping its wings clear of the water to facilitate a quick take-off. It joins concentrations of other terns at spots where fish congregate, such as at warm-water outflows, but is not especially gregarious with other species by nature.

Little and other terns often fish on a rising or falling tide, and rest at high and low tide. They take full advantage of fish shoals that have become trapped in shallow water as the tide retreats. As well as hunting by plunge diving, they pick morsels from the water's surface.

This tern feeds mainly on small fish, including the ubiquitous sandeels, whitebait and other similar-sized fish that occur in shallow water. A small proportion of the diet is made up of other sea life such as small shrimps. Where it nests inland, the bird often hunts over fresh water, where it hawks flies and other insects as well as taking fish.

As with other terns, the male's role in courtship is to bring fish for his prospective mate.

BREEDING

Little Terns pair for life, but have no contact during the winter, so the first task on arriving on their breeding grounds in April is to re-establish the bond with their partner of the previous season. For first-time breeders and those whose mates have not returned, a new partner must be found. By May, established and new pairs alike have completed their courtship. Courting birds display to one another, both on the ground and in mid-air, and the male delivers food to his partner, demonstrating his skill as a provider and also helping the female to achieve breeding condition. The colony as a whole also builds social cohesion, rising as one when a potential predator is around. Sometimes all the birds take off and wheel about together when there is no evident danger – these flights, known as 'dreads', are thought to have a social function.

Both members of each pair prepare a simple nest scrape, usually no closer than 2m to the nearest neighbours. In this scrape the female lays a clutch of typically two or three light stone-coloured, dark-speckled eggs that are well camouflaged against both shingle and sand. The eggs are laid on alternate days and incubation begins as soon as the first egg is laid. The total incubation period is 17–22 days, and usually there is no more than one day between

THE NIGHT WATCH

It is hard to fathom, but illegal egg collecting can still be a serious problem in Britain, and is potentially disastrous when the thieves target Little Terns, which are so susceptible to disturbance (as well as being scarce). In summer 2013 the entire colony at Crimdon, County Durham, was hit by egg thieves who struck at night and raided almost every nest, taking up to 50 eggs.

At Rye Harbour drastic measures have traditionally been taken against such activities, with volunteers recruited to watch over the tern colony through the day and night. Electric fencing has also been used to discourage non-flying predators of all kinds, including humans.

hatches of eggs in the same clutch. The hatchlings are downy and light fawn in colour with darker speckles – the same colour scheme as the egg they just vacated.

Little Tern eggs and chicks are highly vulnerable to predation from both land and aerial hunters. Red Foxes, Hedgehogs, gulls, crows and, in the case of chicks, Kestrels are all significant predators. Flooding from high tides is another hazard that they face. In some years entire colonies fail to rear a single chick. It is not unusual for the whole colony to relocate to an alternative site, perhaps several kilometres away, after a poor breeding season or in response to changes in fish distribution.

The young chicks move away from the nest within a few days of hatching and find separate places to hide, among shady vegetation if available or in hollows in the shingle. Overheating from sun exposure, or chilling if the weather is cold, are both hazards that they face, although the parents brood them or shade them while they are still too young to thermoregulate for themselves. The parent not guarding the nest area will be at sea, hunting for prey. Only very young chicks are fed by regurgitation. Once the chicks are a few days old, the foraging adult brings back one fish at a time, held in its bill, and carefully places it, head first, in the mouth of whichever chick is begging the most insistently.

Chicks are offered whole fish, of an appropriate size for them to swallow without too much difficulty.

The chicks can fly at about 17–22 days. Not long after this they soon begin to catch their own fish and become independent, spending the next month or so honing their fishing skills and gaining weight for the migration ahead, and at this time they may wander some distance from the breeding grounds. They are not usually ready to breed themselves until they are three years old. Ringing recoveries show that Little Terns can live to the age of at least 23. The British longevity record, set in 2013, is almost 18 years.

MOVEMENTS AND MIGRATION

These small birds are long-distance travellers, flying south along the Atlantic coasts of Europe and North Africa to spend their winters exploiting the productive seas off West Africa and possibly much further south. Not all birds stick to the coastal track – some make overland crossings of various lengths, working their way along river valleys. There have been ringing recoveries from inland in France, and also Italy. However, unlike some larger tern species they are not thought to fly across the Sahara.

Migrants depart in August or September, and large numbers may be seen passing coastal headlands as they travel south, feeding on the way. A few may be seen inland, perhaps even feeding or resting at reservoirs or gravel pits – however, this is rare.

THE FUTURE

The Little Tern is without doubt one of Britain and Ireland's most vulnerable breeding species. The extensive beachfront developments up and down our coastline have greatly reduced available nesting habitat for Little Terns, so that the remaining population is concentrated in a relatively small number of 'safe' sites, and a single destructive event such as a spring tidal surge could mean breeding failure for virtually every pair. The concentration also results in larger colonies than might occur naturally, and these can attract a disproportionate amount of attention from predators.

Improved protection for nesting Little Terns led to quite a substantial population rise of 58 per cent between Operation Seafarer (1969–1970) and the Seabird Colony Register (1985–1988). However, there was then a fall of 23 per cent between the Seabird Colony Register and Seabird 2000, which led to the species being placed on the Amber List of species of conservation concern in the UK. There is evidence of a modest increase since Seabird 2000, perhaps a testament to the many ways in which conservation workers have taken action to safeguard colonies and improve their breeding success.

A Little Tern on spring migration makes a stop-off on the coast of Spain.

A mix of adult and juvenile Little Terns on autumn migration take time to rest on an undisturbed beach.

Predation at colonies is a major problem. Land predators can be effectively excluded by fencing off the terns' nesting areas, and this has the added bonus of protecting other vulnerable shingle-nesting species, such as Ringed Plovers, and preventing people from wandering across nesting areas. When erecting fencing it is important to be careful not to disturb the birds, but putting up the fences before the terns return from their wintering grounds may not work, as Little Terns are quite prone to moving around from one spot to another in successive years. One method that has been used to encourage them to a particular spot is to place a few decoy 'dummy' terns, to give the impression that there is already a colony developing there. Playing recordings of courtship calls can also help to attract the birds.

Aerial predators cannot be excluded by fencing. Some of the birds that may take Little Tern chicks are themselves protected and perhaps also vulnerable species – for example, in at least one site in Norfolk, Mediterranean Gulls have been responsible for preying on chicks, and Kestrels have been a problem at others. Supplementary feeding of Kestrels has been tried, but it is unclear whether this has reduced predation. Another way to protect the chicks is to provide shelters for them, in the form of drainpipes sunk into the ground – this offers protection from exposure as well from the keen eyes of gulls and birds of prey.

Flooding is another serious risk to Little Tern colonies. Climate change and associated sea-level rises could obliterate several key sites within the next few decades. At some areas winter work to raise beach levels has been carried out, for example on South Biness island in Hampshire, where the beach was boosted with 500 tons of aggregate in 2013.

Protecting existing nest-sites is one thing, but it would be highly desirable if new areas of suitable breeding habitat could be provided, so that the breeding population could be spread out more and thus be less at risk from 'one-off' events. However, most suitable beaches are heavily used for various recreational activities, and Little Terns could not successfully breed on beaches where people regularly walk unless it was within fenced-off areas. In North America artificial offshore islands have been created to provide extra breeding habitat for terns, with some success. This could be feasible for Little Terns in the British Isles, but creating them would be a major undertaking, with much preparatory study to ensure that they were sited in likely places for colonisation, would not be at high risk of flooding and would not interfere with watercraft navigation.

Fencing is often used to keep walkers away from beaches used by nesting Little Terns and other birds.

Sandwich Tern *Thalasseus sandvicensis*

Cutting a rakish figure with its exceptionally long, slim wings and sizeable dagger of a bill, the Sandwich Tern is a distinctive species and the largest tern to breed in the British Isles. Its population here has fluctuated since the late 20th century, a pattern that seems to be normal for the species. However, it is not a particularly abundant breeding bird here, so it requires monitoring and strong protection to ensure that its numbers stay on a more or less even keel.

INTRODUCTION

This tern is close in size to the Black-headed Gull, but has very different proportions, with a slim body, short neck and legs, and very long, slim wings. The tail is forked but without exaggerated 'streamers', and the wings and back are a very light pearly-grey. With a close view, the long, quite slim bill can be seen to be black apart from the very tip, which is yellow, and the legs are black. The black cap extends into a shaggy crest at the back of the crown, although it looks smooth and flat when the bird is in flight. As the bird moults into winter plumage, the forehead becomes progressively whiter from front to back, like a receding hairline, although the rear crown, complete with the crest, remains black. Juveniles have black scaling on the upperparts, whitish flecks in the crown and solid black bills. The Sandwich Tern has a strong flight with deep, elastic wingbeats. It looks rather front-heavy in flight and often holds its head tilted downwards. It has a very harsh, grating two-syllable call.

This very white tern looks large-headed and long-billed, with very slender long wings.

A flock of Sandwich Terns bathes in the sea close to their breeding colony.

DISTRIBUTION, POPULATION AND HABITAT

Sandwich Terns are widely distributed around the coasts of Britain and Ireland. The most important sites are in north Norfolk, north-east England, north-west Scotland and the east coast of Ireland. Two of the breeding sites in Ireland are inland loughs. The most recent count, Seabird 2000, produced just over 14,000 pairs, some 3,700 of them in Ireland.

This tern occurs on coasts in northern Europe, and around the Mediterranean, Black and Caspian Seas, and migrates to Africa, with birds breeding further east heading to the Arabian Peninsula, southern India and Sri Lanka. The total population is about 69,000–79,000 pairs. The species was formerly considered conspecific with Cabot's Tern (and by some authorities still is), a species found in North and South America.

In the British Isles most Sandwich Terns nest on the coast, on shingle and sandy beaches and rocky shores, as well as on small islands in saline coastal lagoons, and in estuaries. Similar types of shoreline and islands are used by the inland-nesting birds in Ireland. The colonies are often shared with other tern species and also gulls. Sandwich Terns hunt for prey mainly in shallow, calm inshore waters, but sometimes a dozen or more kilometres offshore, depending where the preferred fish species are congregating. When breeding the feeding areas are usually within a 15km radius of the nest-site. However, birds that breed inland also feed at sea, sometimes more than 20km away.

PACKED LUNCH?

You could be forgiven for supposing that the Sandwich Tern is so called because it was a popular bread filling in years gone by. In fact both the bird and the foodstuff are named for the town of Sandwich in East Kent, and the naming of the bird and invention of the sandwich both took place in the 18th century. The sandwich was apparently invented by the 4th Earl of Sandwich, John Montagu, who lived from 1718 to 1792, while the type specimen of Sandwich Tern was taken in the vicinity of the town in 1787.

This colony on the Farne Islands shows how densely Sandwich Terns nest, each sitting bird just beyond pecking distance of its nearest neighbour.

BEHAVIOUR AND DIET

Sandwich Terns are very social, often resting together in small groups when not engaged in feeding or breeding activity. They are also very vocal during the breeding season, even when not close to their nests but flying towards or from the sea. Their standard flight height when 'commuting' is quite high, typically higher than that of other terns. At the colony groups of birds often make their way to the shoreline to bathe in groups, and they often fly in twos or small groups. They seem less inclined to mob predators than other terns do.

Hovering usually precedes a steep plunge-dive into the sea.

Like other terns the Sandwich Tern feeds by surface picking and plunge diving, with the latter method predominating. It is a powerful diver, entering the water from a height of 3m or more in a steeply angled descent with wings drawn back, reminiscent of a miniature Gannet. It may reach 2m under water before surfacing. It is fully immersed for a couple of seconds, but frees itself from the water quite quickly and easily. Even in relatively rough seas it dives frequently, but it needs calm water to feed by surface picking. It will fly into a headwind to help hold its position when flying low and scrutinising the water with its head tilted downwards, looking for prey sufficiently near the surface to be snapped up in flight.

This tern feeds almost entirely on fish, taking any suitably sized species, with clupeids such as herring being well represented in the diet, along with sandeels. In the wintering grounds anchovies and sardines are important. It may also occasionally take swimming crustaceans, squid and marine worms.

BREEDING

Sandwich Terns are the first of our breeding tern species to return to their colonies, with the first birds appearing in March. The location of a colony may change year on year, but it is unusual for Sandwich Terns to strike out alone in establishing a new colony – instead they join existing Black-headed Gull colonies. At new sites the birds may not actually breed for several years.

Some courtship behaviour may already have taken place on the wintering grounds – the earliest arrivals include couples that have been together in previous years. Younger birds without previous breeding experience – those four years old or younger – tend to arrive later. Males without mates

Fish are carried singly, crosswise in the bill.

perform aerial displays, flying high over the colony (usually carrying a fish as added incentive) and calling, and also give their advertising call from the ground and approach other terns in the hope of locating an unpaired female. Once a pair is established the male brings fish for the female, and the offering of the fish is accompanied by ritualised bowing or stretching postures and mutual head-turning displays. Often, copulation takes place when the female adopts a low, begging posture.

The nest, such as it is, starts out as a simple scrape on the ground, and in it the female lays one or two eggs (very occasionally three). There is a gap of up to five days between laying, but incubation begins immediately. Both sexes take their turns at this, and perform abbreviated versions of the courtship display when changing shifts. The sitting bird may steal nesting material from a nearby gull nest, and by the time incubation is well advanced, a fair-sized nest has often developed. Incubation takes 21–29 days, and the chicks may leave the scrape when still quite small, sometimes associating with other chicks in crèches or even on occasion being 'adopted' by another pair – their behaviour seems to depend on the amount of cover available. Usually the parents feed their own chicks within a crèche, and sometimes young birds without young of their own attempt to feed chicks at random.

Both birds in this courting pair are showing traces of winter plumage with their 'receding hairlines'.

NEIGHBOURHOOD WATCH

Black-headed Gulls have a strong community spirit. Their nests are packed closely together and, while immediate neighbours frequently bicker with each other, the whole colony will join forces to see off anything that looks at all predatory, even if that animal was just passing through with no intention of raiding a nest. It takes a very determined hunter indeed to run the gauntlet of several hundred aggressive Black-headed Gulls in full attack mode.

Sandwich Terns are not fiercely defensive, but by nesting among the gulls they get all the benefits of the gull guard, and often they try to squeeze themselves into the most central position possible. The gulls are not perfect neighbours. They try to steal fish from arriving terns, and may even help themselves to a tern egg or chick on occasion. However, it seems that for the terns this is a sacrifice worth making, as the majority of our Sandwich Terns do breed among Black-headed Gull colonies.

Young birds in juvenile plumage have distinctly scaly upperparts and a general brownish tinge.

The chicks can fly at about 30 days old, and they and their parents leave the colony soon afterwards. The young birds continue to be fed by at least one parent for an extended period afterwards – up to three months – while the young terns gradually improve their own fishing skills. Those that survive their first winter stand a good chance of living a long life – the record is nearly 31 years old. Like other long-lived birds with low annual productivity, they often forgo breeding for a year or more when their condition is suboptimal, or when there are problems on their breeding site.

MOVEMENTS AND MIGRATION

Dispersal from the colonies takes place throughout July and August, and may be in any direction, with some birds heading north-east as far as Sweden and Poland. By September nearly all birds – adults and youngsters alike – are on their way south, the adults already in winter plumage, and most are gone by October. They start to reach their wintering grounds in November. Sightings and ringing recoveries suggest that the majority travel along the Atlantic coast, some wintering in West Africa and others down to the southern tip of the continent from Namibia round to Mozambique. However, some migrants take overland shortcuts; a few British-ringed birds have been recovered well inland in eastern Europe, others in the Mediterranean. Older birds overwinter further north, while first-year birds remain on the wintering grounds throughout the following summer. On their wintering grounds they feed in inshore seas, and also in estuaries and river mouths, but only occasionally inland. They are very social in the wintering areas, roosting in large groups, often with other tern species.

A tiny number of Sandwich Terns overwinter on British coasts. These are likely to be birds that breed further north, and are most likely to be encountered in western Ireland or south-west England.

THE FUTURE

This charismatic tern has had mixed fortunes in Britain and Ireland in recent years. Numbers increased by 33 per cent between Operation Seafarer and the Seabird Colony Register, but then fell by 15 per cent between the Seabird Colony Register and Seabird 2000. The population since 2000 seems to have remained fairly stable. This overall picture masks some more marked regional variations. There has been a very steep decline in Scotland (of 53 per cent since the Seabird Colony Register in the mid-1980s), and a much more modest fall in England (8 per cent since the Seabird Colony Register) while in Ireland numbers have risen considerably (68 per cent since Operation Seafarer in 1969–1970).

Because they depend on good supplies of fish of a certain size within a fairly limited foraging range, these terns can suffer breeding failure if fish stocks are low. Studies in the Netherlands have shown that the productivity of Sandwich Terns by the Wadden Sea was closely correlated with catches of clupeid fish by fisheries. In years when fewer fish are available the terns suffer not just from a shortage of prey, but from having some of their catches stolen by their gull neighbours. A certain rate of such kleptoparasitism occurs each year, but this usually has no particular impact on breeding success in normal years. However, it can make the difference between survival and starvation for chicks in years of poor fish supplies. Early in the season small sandeels are more important than clupeids as food for chicks.

Wherever they travel Sandwich Terns need fish to eat, and in the wintering grounds there are marked year-on-year variations in numbers and distribution of suitable fish such as anchovies and sardines. First-year survival rates of Sandwich Terns are fairly low at 45 per cent (compared with 82 per cent for adults), and some of this will be down to birds failing to make the grade when fishing for themselves through the winter. There is also a tradition of trapping terns in parts of West Africa, which may have considerable impacts on winter survival rates.

Breeding birds can suffer high predation, especially by Red Foxes. If a fox discovers a colony the result is total breeding failure, because the mammal will predate every nest over the course of a few nights. Colonies on nature reserves are protected from these and other land mammals by electric fencing – this must of course be carefully monitored and quickly repaired if damaged. Predation by birds is less serious a problem for the terns, partly because of the robust defences of their gull neighbours. However, if there are too many gulls they can make it impossible for terns to nest among them – they will outcompete the terns for space, and pressure in the form of kleptoparasitism and predation of tern eggs and chicks may be too severe. Disturbance by humans, however accidental or benignly meant, can also cause an entire colony to abandon its breeding attempt.

Sandwich Tern has declined severely in the most northerly parts of its British range.

The effects of severe weather can be devastating for nesting terns, with heavy spring rain and low temperatures placing chicks at risk of chilling, and exceptionally high tides liable to flood out nest-sites. Bad weather in the winter can also have a serious impact on breeding sites, by rearranging the beach so much that it becomes unusable for the terns when they return. Artificially raising beach levels can help, as can providing safer sites in the form of islands on coastal lakes and lagoons.

Sandwich Tern conservation requires quite a high-intervention approach and a lot of creativity, to try to keep their beaches safe for them. At the same time great care must be taken to minimise disturbance. Managing fish stocks is more difficult and uncertain, and there are probably natural fluctuations at work alongside the possible effects of overfishing and human-caused climate change. Increasing the number of colonies will help to provide overall safeguarding against local declines caused by these fluctuations.

A handful of Sandwich Terns overwinter in the British Isles – perhaps climate change will see more adopting this strategy.

Roseate Tern
Sterna dougallii

The Roseate Tern is our rarest native seabird, with the bulk of its population in Ireland. It is a particularly beautiful tern, with the longest tail streamers of all our species. Following a long-term decline, efforts to improve its fortunes in Britain have paid dividends, with a strong population increase since 2000, but it is still very uncommon and vulnerable, with a Red conservation status in the UK.

INTRODUCTION

This tern is about the same size as a Common Tern, but has subtly different proportions – it is longer-tailed and shorter-winged, with shorter legs and a slightly longer and thicker bill. Its wings and back are a very pale grey, comparable to that of the Sandwich Tern, with the outer primaries being slightly darker. The bill is black with a hint of red at the base, and the legs and feet are red. Some birds show a delicate rosy flush on the breast and belly, although others look clean white. The black cap recedes from the forehead in winter. On the wing the Roseate Tern has a fast-flapping flight reminiscent of a Kestrel or other small falcon's. Its calls are loud and harsh. Juveniles have dark-scaled plumage with very little brown in the wings compared with Common Tern juveniles, and short tails.

Rarest of our 'sea terns', the Roseate Tern is rather long-legged and very long-tailed, with almost white upperparts.

A Roseate Tern flying alongside a Sandwich Tern (right) illustrates the size difference between these two dark-billed species.

DISTRIBUTION, POPULATION AND HABITAT

The main sites for breeding Roseate Terns in the British Isles are Coquet Island off Northumberland, Lady's Island Lake in County Wexford, and (by far the largest colony) Rockabill island (actually a pair of islands) off County Dublin. Other colonies are tiny and some are only sporadically occupied. There are in the region of 90 pairs in the UK and more than 1,000 in Ireland – although numbers can vary considerably from year to year. The birds migrate to West Africa for the winter.

The colony on Rockabill is the largest in the whole of Europe – the total European population is just over 2,000 pairs. The Azores hold the most breeding birds after Ireland, with much smaller numbers in France and Spain. The species is quite widely distributed beyond Europe, breeding in North America and the Caribbean, southern and East Africa, the Seychelles, islands in the Indian and Pacific Oceans, Japan, Taiwan and Australia, which holds the bulk of the global population – 83,000 pairs out of 120,000–130,000 worldwide.

Roseate Terns nest on rocky or flat shorelines or on higher ground when on small islands, often among vegetation and sometimes alongside other tern species, especially Common Terns. They feed in inshore seas and very rarely occur inland.

BEHAVIOUR AND DIET

Rather less social than other terns, Roseates are fairly often seen hunting or resting alone as well as in small groups. If seen flying among Common Terns they show a stiffer-winged, more powerful flight action. When foraging they usually stay within 10km of their colonies.

Roseates are strong plunge divers, entering the water from a height of 2–3m. Before the plunge they may fly into the wind and hover to hold their position while checking the movements of fish in the water below. The dive may take them a metre or two under water. They may be attracted to other feeding seabirds, revealing the presence of a shoal of small fish. They also feed by surface picking in calm conditions, and sometimes steal prey from other tern species.

The diet consists almost entirely of small fish, with sandeels, various clupeids and also gobies making up the bulk of it during the breeding season, in varying proportions depending on time and availability. Alternative prey such as crustaceans is taken only occasionally; Common Terns in the same colony take a much more varied diet than do Roseates.

A lone female Roseate Tern at a Merseyside nature reserve persuades a male Common Tern to offer her a fish.

BREEDING

This tern is a late breeder that is not usually back on its breeding grounds until late May to begin nesting in June. Only birds three years old or older breed, although two-year-olds do return from Africa and may visit colonies. More than 90 per cent of surviving adults that have bred before return to the same colony in subsequent years, and this becomes increasingly likely with age. They are also very likely to pair with the same partner as that of previous years. In some years pairs make no breeding attempt; this is most likely if conditions on the nesting grounds have changed considerably over winter.

The courtship period involves the pair flying high together over the colony and calling, the male usually with a fish in his bill. Unpaired males perform the same flight and with luck are soon followed by an interested female, or perhaps more than one. Once pairs are bonded and the female is getting ready to lay, the male regularly brings fish for his mate.

The nest-site is better hidden than the sites of other *Sterna* terns, and is usually near or under some sheltering rocky overhang or concealed in vegetation or debris; it is a shallow scrape, dug out by both parents. In it the female lays one or two eggs, and she carries out the bulk of the 21–26-day incubation, the male relieving her briefly during the day. There is a two- or three-day gap between laying in two-egg clutches, so one chick is clearly older than the other. Studies have shown that in broods of two, once the chicks are flying the parents divide their efforts consistently between the chicks, the male feeding the older chick and the female the younger. However, it is rather common for the younger chick to starve at an early age.

The chicks leave the nest when still small and hide separately in vegetation or other shelter. This helps them to avoid being spotted by predators, although on rocky shores they run the risk of falling into gaps between rocks. Both adults hunt fish for the chicks, offering progressively larger specimens as the chicks grow larger. The chicks begin to roam more widely around the colony when they are about three weeks old, and within another ten days they are taking their first flights. Once able to fly each chick leaves the colony, accompanied by one of its parents if it is in a brood of two, while the other stays with the remaining chick until it, too, can fly. Parental care continues for another six weeks or so.

MOVEMENTS AND MIGRATION

Adults and young birds disperse from their colonies in August. They head in various directions, including north-east to the coasts of Belgium, Holland and Germany, but throughout September they leave Europe and set off on their journey south. At this time they may be seen offshore from headlands, although most go by well offshore. They proceed quickly southwards along the coasts of France, Spain and North Africa, to a relatively small area of coastline (mainly Ghana) off West Africa where they spend the winter. A few cross the Atlantic instead – one Rockabill-ringed chick was found in Brazil in January 2001, and two other Irish birds have reached the USA.

THE FUTURE

The Roseate Tern colony on Rockabill has shown a dramatic population growth. Just 60 pairs were present in 1969–1970, but by the mid-1980s this figure had increased to 227 pairs, and Seabird 2000 found 618. Since 2000 numbers there have sometimes exceeded 1,000 pairs. This is in stark contrast to the species' fortunes elsewhere in the British Isles. Between Operation Seafarer (1969–70) and the Seabird Colony Register (1985–88), numbers in England fell from 355 to 34, and have shown only a modest recovery since then. The Scottish breeding population has fallen by a similar magnitude, as have other colonies in Ireland.

Perhaps the obvious explanation for this is that birds have deserted other colonies and decamped to Rockabill, and it is likely that this has happened to some extent. However, that would still leave many Roseate Terns unaccounted for, with an overall decline of 67 per cent between Operation Seafarer and Seabird 2000. Some large colonies have disappeared altogether – the only other site to show encouraging signs of recovery is Lady's Island Lake in County Wexford, where a colony has become re-established following its disappearance between the 1970s and 1990s. However, this colony is still a shadow of its former self, comprising only about 100 pairs, compared with 1,352 pairs present during Operation Seafarer.

The possible causes of the Roseate Tern's declines are multiple. High chick mortality because of inclement weather, starvation and predation are all factors. On the wintering grounds many first-year birds were recorded in the 1980s to be trapped and killed by locals. A high proportion of young birds spend their first year of life off Ghana, where the practice of hunting terns is possibly still widespread. Availability of fish in the seas there is also highly variable, and that is unlikely to improve as industrial fishing fleets from other nations are now exploiting the area. High breeding productivity is necessary to offset the resultant low survival of young and wintering terns.

The successful site of Rockabill has a number of characteristics that make it a good breeding site for Roseate Terns. The island that holds the breeding terns ('The Rock') is free of predators and uninhabited, with no public access, and is managed by BirdWatch Ireland as a nature reserve, with a particular emphasis on safeguarding and supporting the Roseate Terns. The birds are provided with an abundance of nestboxes, which are placed in suitable spots among the rocks and protect the chicks from bad weather and predators. Nestboxes have also been set up on Coquet Island, Northumberland, the one nesting site for the species in a county that once held more than 300 pairs, including on the Farne Islands and Lindisfarne.

Only very proactive measures will safeguard this beautiful tern's future as a British breeding bird.

To hang on to the Roseate Tern as a breeding species, we need to continue to support the breeding pairs that remain, but also to do all we can to improve their chances of survival over winter. The species is rare and declining across Europe, so there is rather limited potential for recolonisation if more colonies are lost.

Common Tern
Sterna hirundo

The scientific name *hirundo* means 'swallow', suggesting that this tern is the original 'sea swallow'. It is the most common of our terns over much of the British Isles (in fact Arctic Terns are more numerous, but are much more restricted in their distribution), and is also the tern most likely to be found breeding inland (although most colonies are on the coast). Even though it seems to be more adaptable than our other terns, this species is in decline in the British Isles. It is quite responsive to conservation efforts, in particular provisions for it to colonise lakes and reservoirs inland, but protecting the coastal colonies is more difficult.

INTRODUCTION

This is a graceful long-winged and long-tailed tern. It has a mid-grey back, rump and wings, with distinctly darker grey on the outer primaries, forming a 'wedge'. On adults in breeding plumage the neatly defined cap is black, the bill is red with a black tip, and the legs and feet are red. A few individuals have much blacker bills and could be mistaken for Roseate Terns, but they have much darker upperparts than that species. The underside is whitish but not usually perfectly white – it tends to look slightly greyish alongside Black-headed Gulls and Sandwich Terns. In winter the forehead becomes white, and the bill and legs turn darker. Juveniles have scaly brown markings on their backs and wings, and shorter tails without streamers on the outer tail feathers. In flight this species looks very buoyant and graceful. Its call is a loud, shrill and rather grating two-syllable screech.

The black bill-tip and diffuse dusky edge to the wing helps separate the Common Tern from the very similar Arctic Tern.

This is our only tern species which regularly breeds well inland.

DISTRIBUTION, POPULATION AND HABITAT

Common Terns have a wide distribution around the coast of the British Isles, with major concentrations on the south coast of England, East Anglia, north-east England, north-east and western Scotland, around Liverpool Bay, North Wales and the east coast of Ireland. They do not breed around the south-west coast of England and most of Wales. There are strong inland populations in the Home Counties and East Midlands, and a few others in central Scotland and Ireland. The total breeding population for the UK is about 12,000, with another 2,500 in Ireland.

This tern is widespread globally, breeding across Europe and Asia, and also North America, although it avoids the most northerly coastlines. All populations migrate south for the winter, so over winter Common Terns can be found on coasts around South America, South and South-east Asia, Australia and Africa down to the southernmost tip. The total world population is 1.6–4.6 million individuals.

COMIC TERN

Separating the Arctic Tern from the Common Tern is difficult unless views are very clear and close, but there are many subtle differences to look out for, which are summarised here:

- **Overall size and shape** Common is larger but shorter tailed than Arctic.
- **Bill shape** Bill is longer in Common, while Arctic's is shorter and looks stouter and straighter.
- **Bill colour** Red with a black tip in Common, and slightly pinker red with no black tip in Arctic.
- **Leg length** Longer legs in Common. Arctic looks closer to the ground.
- **Underside colour** Almost white in Common, clearly grey in Arctic, often with distinctly whiter cheeks.
- **Wing pattern** Common has a broader and more diffuse dark border to the 'hand' of the wing – in Arctic the dark border is narrow and crisply defined.
- **Wing translucency** Against the light Common shows a translucent area in the primaries, but in Arctic all the primaries and secondaries look translucent.
- **Juvenile plumage** Common looks quite brown on the upperside; Arctic appears much greyer.
- **Moult** Common begins to moult its primary feathers before migration, while Arctic moults on its wintering grounds. Hence if a bird is showing wing moult in summer or autumn it should be a Common.

Location can also help with identification – very few Arctic Terns breed in England, for example. There are many areas in Scotland and Ireland where both species breed, and can be compared side by side. However, poorly seen distant birds may be unidentifiable, so go into the birding notebook as 'comic terns'.

Hovering with the head tilted down, a Common Tern is poised to dive after prey.

Common Terns breed on flat beaches, both sand and shingle, and on rocky shorelines. They often share their colonies with other terns and sometimes small gull species. They also nest on shingly and lightly vegetated islands in lagoons, lakes and reservoirs, and readily adopt artificial floating 'tern rafts' with a scrape-able substrate placed on top. They fish at sea and also on rivers, canals, pools and lakes.

BEHAVIOUR AND DIET

Common Terns are quite social and are frequently seen resting in small groups or flying together over water. They are highly aerial – with a large wing area relative to their body weight they can stay airborne in active flapping flight with relatively low energy costs. When overflying potential feeding areas they tilt their heads downwards and fly slowly, sometimes hovering or making knife-edge turns to double-check the water directly below them. They follow fish shoals moving under water until the fish come close to the surface, or swim into water shallow enough for the terns to access them.

The birds feed by plunge diving, usually not from a great height, and often do not fully immerse in the dive but splash their bodies down while keeping their wings clear of the water. A short hover often precedes a dive. They also feed by surface picking, and may catch insects in flight. They usually forage no further than 20km beyond the nesting area.

This tern is a more adventurous feeder than most other *Sterna* species. However, its diet still comprises primarily small fish (up to 15cm long). Most of these are clupeids such as young herring and sprats, as well as some sandeels. Terns foraging over inland waters take a high proportion of sticklebacks, as well as immature fish of larger species. The remainder of the diet is made up of crustaceans and insects (both swimming and flying), and occasionally squid and floating carrion.

As well as plunge diving, Common Terns feed by picking floating morsels from the surface.

Tern rafts have high sides to protect nests from waves and discourage mink.

BREEDING

Like our other terns this is a summer visitor, and most birds arrive during April. For the first few days they spend much time overflying the nesting area and calling, but gradually as numbers build they spend more time on the ground, the males competing for the best spots to establish their nesting scrapes, and attempting to attract females. Males also fly over the territory carrying a fish, and walk among the other terns with a fish, in the hope of finding an unpaired female. The chance of a pair reforming in successive years is high and increases with age, as does the likelihood of the birds using the same territory as in previous years. Both members of a pair drive away other terns that encroach onto their territory.

Courtship feeding is a major part of pair-bonding behaviour. The female initiates this by begging in a crouched posture, like a fledgling – this is the male's cue to go fishing. In due course he returns to his mate with a fish in his bill, calling loudly, and that may trigger her to take off and fly after him, or the fish may be passed over immediately. By feeding the female and allowing her to rest rather than forage for herself, the male can help his partner achieve peak condition for egg laying, which takes place in May. The displays that precede copulation do not necessarily involve fish, but the birds circle each other on the ground in an exaggerated posture with wings half open. Often the male jumps on the female's back a few times without actually attempting to copulate, before an actual mating takes place.

FISHING RIGHTS

Most seabirds do not defend individual territories on their feeding grounds, because they range too widely (in response to the variable distribution of their prey) for there to be any advantage in defending one area from others of their species. Terns, however, do sometimes have feeding territories. Common Terns in North America have been found to use specific parts of coastline during the breeding season. Each territory is 150–300m long and extends up to 75m from the shoreline. Both members of a particular breeding pair use the same patch and defend it from other Common Terns.

This behaviour was noted along a particularly complex stretch of coastline with many bays and estuaries, making the absolute length of coastline proportionately long and allowing space for territorial behaviour. On straighter coastlines such behaviour has not been observed.

Nestboxes offer some weather protection for nesting terns, although they do not necessarily choose to use them.

The clutch is usually of two or three eggs, laid on successive days or at two-day intervals. Incubation begins straightaway in many cases, but the birds do not start to 'sit tight' until the second or third egg has been laid, so that the interval between hatching times is shorter than the interval between laying times. The incubation period is 21–25 days and both birds share the task, taking alternate shifts of between three and five hours, with the off-duty bird away foraging by day and resting nearby at night. The birds also gradually build up the edges of the nesting scrape during the incubation period, adding pebbles and vegetation to the margin to form a more distinct nest.

The chicks are downy with open eyes on hatching, and become quite mobile early on, usually leaving the scrape at three or four days old to shelter in vegetation nearby. At this point the terns' responsiveness to potential predators – always quite vigorous – goes up a notch, with all birds rising up to mob a passing gull or human. The chicks protect themselves by hiding, or by freezing and relying on their camouflage to conceal them.

The adults feed the chicks at the nest-site for 22–28 days, when the young birds take their first flights. Then adults and chicks both leave the colony, but the parents continue to feed the chicks for another month or two while the young birds learn how to find and catch fish of their own. Young birds spend their first full year off Africa, returning to their colonies as two-year-olds, but most do not breed until at least four years old. Those that survive their first year can live a long life – the longevity record is 33 years.

SOCIAL STRIFE

The distance between two Common Tern nests within a colony varies considerably depending on how much suitable space is available and how it is distributed. In some cases nests are less than a metre apart, and there can be considerable stresses involved with living in such close proximity, especially once the chicks become mobile.

Terns are aggressive to each other when defending their territory, and that aggression is not limited to adult birds; youngsters that wander into their next-door neighbours' territory may find themselves fiercely attacked and perhaps even killed. This might seem extreme, but there are good reasons why it takes place. Young Common Terns regularly practise kleptoparasitism, snatching freshly delivered fish from the bills of other chicks. If they are larger than their victims they are very likely to succeed in this. It is therefore sensible for terns to keep other pairs' chicks away from their own.

MOVEMENTS AND MIGRATION

All of our Common Terns migrate to Africa, but those that breed further north go further south, reaching south of the Equator, while more southerly breeders winter off West Africa or even Iberia. The migration begins in September, but first the birds spend some time dispersing in various directions from their breeding grounds, some going as far north-east as Finland. In due course the whole population begins migration proper in a southerly direction, and arrives at the wintering areas through November. Many birds from further east in Europe pass British coasts on their southbound migration.

Ringing recoveries of British-hatched chicks show that the birds mainly move along the Atlantic coast, although some go overland, and that the wintering population is concentrated around Senegal and Ghana, but ranges to South Africa. One exceptional and baffling record concerns a bird ringed as an adult in County Down in 1959, which was found dead in south-eastern Australia nine years later, apparently in the wake of a severe storm.

In juvenile plumage, Common Terns have distinct brown or ginger tones to their plumage.

THE FUTURE

After showing relative stability over surveys in 1969–1970 (Operation Seafarer), 1985–1988 (Seabird Colony Register) and 1998–2002 (Seabird 2000), the Common Tern population fell by about 24 per cent in 2000–2012.

Predation by mammals, especially Red Foxes and American Mink, is one of the main problems that Common Terns face, and the minks with their strong swimming abilities are able to reach some island colonies as well as those on the mainland. Help in this department may be coming from an unexpected source – the Otter is increasing and spreading throughout the British Isles, and seems to outcompete the non-native mink in all kinds of habitats. Common Terns that nest on nature reserves can be protected from predators by fencing, and on some inland reserves minks are trapped and culled.

Common Terns that feed in the North Sea have faced the same issues related to fish shortages as other seabirds. They can adjust to a certain extent as they do take a relatively varied diet, but in years of low fish supply the northerly populations suffer more nest predation from skuas and gulls. They may also be outcompeted for nest-sites by nesting gulls. Habitat loss is another factor, with progressive disturbance or erosion making some sites unsuitable. Like all species that nest on beaches just over the high-tide line, they can lose their nests to an exceptionally high tide, and in the future may lose breeding areas as a consequence of predicted sea-level rises. Building up beach levels by adding shingle or aggregate can help to offset this.

Trapping by humans on the wintering grounds is a problem faced by this species and other British terns. In the recent past, the practice is common in Ghana in particular and seems to be a children's pastime, done as much for sport as to use the terns for food. Local education aimed at discouraging the practice has had some effect at reducing trapping levels, but this needs to be consistently maintained.

Common Terns are gratifyingly easy to encourage at inland sites, and many English nature reserves with open water have used tern rafts to great effect. Rafts are relatively cheap to make, and their design is improving year on year as the needs of the nesting birds and their chicks become more fully understood. However, most sites are not large enough for the amount of rafts that would be needed to support more than a couple of dozen pairs of terns. A network of small inland colonies provides good security for the species as a whole, but the large coastal colonies are in a vulnerable position and in need of strong proactive conservation management.

Arctic Tern
Sterna paradisaea

Seabirds have a habit of getting themselves into the record books. The holders of avian records for the largest wingspan, longest lifespan in the wild, deepest and longest dives, and the best ability to endure extreme conditions – are all seabirds. The Arctic Tern earns its place in the hall of fame for an annual migration of almost legendary proportions. Individuals that nest furthest north migrate the furthest south, and therefore see constant daylight for most of the year. In the British Isles the Arctic Tern has suffered the familiar pattern of increase throughout the early 20th century, followed by a sharp decline towards the end of the century as the essential sandeels and other fish became difficult to find. It is still, however, our most abundant tern species. A visit to a colony in full swing is an unforgettable experience – and one that necessitates some head protection for the visitor, because these terns are notoriously fierce in defence of their nests.

INTRODUCTION

The Arctic Tern is a medium-sized tern, just a shade smaller than the Common Tern but with more elongated tail streamers. When perched it looks more monotone, with its greyish underside contrasting little with the colour of the wings and mantle. Overall it has a squatter shape than that of the Common Tern, with shorter legs and neck. Its legs and bill are both crimson, the bill is relatively short and straight, and the cheeks can look very white against the greyer neck-sides. In flight it looks very light, and has a less purposeful and more 'bouncy' action than that of the Common Tern. It shows a neat, narrow black border to the 'hand' of the wing on both the upperside and underside. Juveniles have black scaling on their grey uppersides and a white forehead. The call is a disyllabic, shrill and harsh 'keee-arrr'.

A bird of tremendous grace and beauty, the Arctic Tern has endurance that belies its slender frame.

Their short legs make Arctic Terns awkward on land, so they are always ready to launch skywards should they need to see off a predator or a rival.

DISTRIBUTION, POPULATION AND HABITAT

As befits its name, the Arctic Tern is found mainly in the north. Scotland holds the vast majority of our breeding pairs, and Orkney and Shetland are home to most of the Scottish population. There are also significant colonies on the Western Isles, on islands off Northumberland and on Anglesey (the most southerly colonies of any size), and there are several modest-sized colonies around the Irish coast. The most recent counts suggest that there are about 47,000 pairs in Scotland (7,300 on Shetland and 32,500 on Orkney), 3,600 in England (almost all in Northumberland), 1,700 in Wales (all on islands off Anglesey) and 3,500 in Ireland. The total count of just over 56,000 found by the Seabird 2000 survey is thought to have declined by about 5 per cent in 2000–2012.

This tern breeds around the entire Arctic and sub-Arctic, and southwards it reaches as far as Brittany in France, and Massachusetts in North America. All birds migrate southwards for winter, some reaching the Antarctic continent. The world population is estimated at 2 million individuals.

The majority of Arctic Terns nest on coasts, although in some parts of their range some birds breed inland on tundra in the High Arctic. They use all kinds of shoreline, including shingle, sand and rocky ground, and also nest on islands within coastal lagoons, on rocky islets and on shingle spits. They typically feed in inshore seas, within 3km of the colony when breeding, but inland breeders hunt over boggy pools in the tundra. Breeding colonies are often shared with other tern species, and in the far south of the British Isles it is sometimes possible to track down one or two pairs within a large Common Tern colony. The two species have hybridised in the wild.

On migration the Arctic Tern crosses open sea, although it may come close to the coastline or rest on beaches – it will also rest on boats and other floating objects. Some individuals make overland crossings while migrating. In winter the birds are more strictly marine than when breeding.

The all-scarlet, relatively short bill helps to identify this as an Arctic Tern.

Arctic Terns do not plunge deeply for prey, and dives are often little more than a bellyflop.

BEHAVIOUR AND DIET

This tern is very gregarious, both when breeding and while fishing, and also on migration. It has a very buoyant and rather jerky flight, which is more hesitant and involves more frequent hovering than the flights of Common or Roseate Terns.

Like other terns the Arctic Tern has a steady and slowly progressing hunting flight – it travels a few metres above the water and keeps its head tilted down to check for fish. It makes frequent dips and turns, and may hover, drop a short distance and hover again a few times before actually plunging, with the dive being powered by wingbeats as well as gravity. The touchdown on the water is often a bellyflop rather than a full dive, with the wings staying clear of the surface. Arctic Terns also pick floating food items from the water's surface, and chase flying insects. They have been known to occasionally feed on carrion.

The birds are skilled at finding areas where suitably sized fish are abundant, and follow shoals as they move along, hoping for an opportunity to catch them as they move into relatively shallow water. On their wintering grounds they are known to sometimes associate with feeding whales, taking fish that are being driven towards the surface as the whales move. Similar associations may occur with large predatory fish. On inland breeding sites in high tundra, insect prey can be quite important.

The diet is mainly of fish, with occasional crustaceans, insects and other invertebrates. Sandeels form an important part of the diet, especially during the breeding season, although in some areas they are more important than in others. Studies on Anglesey's island colonies found that nearly 100 per cent of the diet consisted of sandeels, but on the Farne Islands they made up only 22 per cent, with clupeid fishes like sprats, herring and Saithe making up the bulk of the diet. They exploit mass emergences of insects from the surface of fresh water, and in some American colonies shrimps are more important prey than fish.

Growing chicks require deliveries of ever larger fish, each of which will be swallowed whole.

MIND THE MOB

One of the essential destinations for any birder keen to get up close and personal with seabirds is the Farne Islands, off Northumberland. This group of up to 20 islands (some disappear at high tide) is protected by the National Trust and holds important colonies of many seabird species, including the Puffin, Eider and Arctic Tern. Organised boat trips visit the islands in the breeding season, and those disembarking on Inner Farne can expect to see all of these species and more.

However, the trip is not without its dangers. The Arctic Terns respond with their trademark aggression to the arrival of humans within their colony, and they do not hold back when they attack. Anyone unwise enough to arrive hat-less is liable to end up with a bleeding scalp from repeated tern pecks. On the other hand, the birds' swooping attacks make for exceptional photographic opportunities.

Understandably, concerns are often raised as to whether the visitors' presence is harmful to the terns, which after all must be somewhat stressed by the perceived danger and, in mobbing people, are leaving their nests unattended and perhaps exposed to genuine danger – large gulls live on the islands and sometimes predate tern nests. Studies have been carried out to compare hatching times in nests without disturbance with those where the terns take off to mob visitors during the 3½-hour island tours. These have found that incubation takes on average a day longer among the nests subject to disturbance. However, this is offset by reduced gull predation – the visitors' presence actually reduces the risk of nests being attacked, as they keep the gulls away. The situation on the Farnes is closely monitored by the islands' wardens, and it has been found that more and more terns are opting to nest close to the walkway used by the visitors. They still engage in mobbing people, but on some level they seem to recognise that humans pose a much less serious risk than gulls.

BREEDING

Arctic Terns return to their colonies in late April or May, and quickly establish territories and partnerships. Most birds return to the same colony each year, but relocating to another one nearby is not unusual. As is the case with the Common Tern, the early stages of courtship are characterised by the 'fish flight', in which a male flies high over the colony, calling loudly, while carrying a fish. Interested females pursue him in a dramatic aerial chase, although both sexes preferentially pair with the mate they had the year before. The actual handover of the fish takes place on the ground, both birds adopting a curious stiff posture with wings partly opened as they strut around each other before the male passes over the fish. As courtship progresses these fish exchanges become less about display and more about straightforward provisioning.

The female lays one or two eggs in the nest scrape, and both parents keep them warm for the 20–24-day incubation period, taking turns through the day and night. Most pairs in the colony are on eggs by the end of May. Both eggs and young have cryptic, speckled brown colours, providing

A tiny fish for a tiny chick. This youngster still has its egg-tooth, the hard knob on the bill-tip that helped it to break out of its egg.

camouflage that comes into its own when the chicks are a few days old and leave the scrape, hiding among whatever nearby cover is available. The adults make frequent trips out to sea to fish, and bring back prey one item at a time. They must dodge the attentions of Arctic Skuas in some areas, but many also lose their catches to other Arctic Terns, and chicks also snatch food from each other. Adults steadily lose weight and body condition throughout the period of chick rearing, and birds that suffer the most in this respect also tend to be those that lose one or both chicks to starvation.

All being well the young terns are capable of flight at 21–24 days old. They and the adults leave the nesting colonies soon afterwards but stay together, the parents feeding the chicks for at least another month. By the time migration commences the chicks are independent, catching fish for themselves, albeit with a lower success rate than adults. Young birds stay on the wintering grounds until they are two years old, honing their skills and building body condition, and make their first breeding attempts at three or four years old.

MOVEMENTS AND MIGRATION

Each year every Arctic Tern makes a round trip of 40,000 to more than 80,000km, from the far north of the planet to the far south. The birds set off through July to September, and may disperse north or east before switching to migration mode and heading south. Ringing recoveries indicate that they track down the coasts of western Europe and West Africa to winter around the southern tip of South Africa. However, several ringed birds have ended up in southern Australia and one was recovered at sea in the Southern Ocean, much closer to Antarctica than to South Africa. Two second-year birds from the Farne Islands were found in Russia, and a bird ringed as a chick in Shetland was found in the USA 14 years later, showing that not all Arctic Terns return to their natal colony (or anywhere near it) when they reach breeding age.

The juvenile plumage of this tern is very grey-toned, unlike the brownish Common Tern juvenile.

SECRETS OF THE SEAFARERS

In 2007 researchers trapped a number of adult Arctic Terns that were nesting on Iceland and Greenland, and fitted them with trackers. Those that returned the following year were recaught and their trackers were recovered. These were small, simple geolocater devices that measured light intensity, allowing the researchers to examine day length and sunrise and sunset times. This was sufficient data to provide a picture of where the terns were at any given point in their migration. The results showed that the birds migrated south well out at sea as far as West Africa, then either continued down along the African coast or crossed over to move south along the coast of South America. The return journey, well out to sea, described an S shape, moving closer to southern Africa at first, then crossing back to complete the journey off the coast of Brazil and North America.

As well as providing a detailed picture of the route taken, the study revealed places where the terns stopped off to refuel in between journey 'legs'. These are highly productive areas at sea, used by the terns to prepare for migration and, on the way home, to build their condition ready for breeding. Identifying areas like these is key to providing effective conservation for seabirds outside the breeding season.

THE FUTURE

The Arctic Tern is a high-profile victim of the well-documented declines in sandeels in the North Sea. To what extent the sandeel problems are caused by overfishing, climate change or natural redistribution is unclear. A shortage affects Arctic Terns directly, by reducing the amount of food available for them and their chicks, and indirectly, by increasing the likelihood of skua predation (although this is only a problem in northern Scotland), and by obliging the adults to spend more time away from their nests, leaving the eggs and chicks at risk from chilling.

Like many other seabirds the Arctic Tern showed quite a strong population increase between the Operation Seafarer survey in 1969–1970 and the Seabird Colony Register in 1985–1988, rising by 50 per cent overall, with the most marked increases being in Scotland and Ireland. Then, between the Seabird Colony Register and Seabird 2000, numbers fell by 31 per cent – most markedly in Scotland and England this time. Numbers in Wales and Ireland have increased across all three surveys, although in Ireland there have been losses of many small colonies. There are concerns over concentration of the bulk of the population at a shrinking number of sites, especially as this species naturally tends to form more small than large colonies.

Travelling from pole to pole each year, some Arctic Terns only encounter twilight and night when on migration.

Besides the problems of food shortages (and related skua predation), Arctic Terns have been impacted by the introduced American Mink, a voracious nest predator and keen swimmer to boot. Mink predation has been particularly severe in the rugged north-west of Scotland – tern colonies in more populous areas are likely to be on nature reserves, where mink control tends to be practised. Adverse weather can also affect breeding success. On the positive side, Arctic Terns are likely to be safer on their remote wintering grounds than terns that overwinter further north.

The main threat facing this species is without doubt food-supply issues, so identifying and protecting key feeding areas is key to ensuring the continued survival of this exceptional and inspiring bird.

Other terns

All three species of Eurasian 'marsh tern' have been recorded in Britain, but only one – the **Black Tern *Chlidonias niger*** – is regularly seen offshore. However, it is really a land bird and is probably more likely to be encountered over a lake or reservoir than on the coast. This attractive small tern was once a British breeding bird, but since widescale drainage of the Fens and other suitably marshy landscapes it now only makes sporadic breeding attempts.

Black Terns are widespread in Europe, although their distribution is very patchy in western Europe. There are colonies close to the British Isles, in France, Belgium and the Netherlands, and a hundred or more migrants are recorded in Britain every year – sometimes many more. The adult in breeding plumage is very distinctive with its sooty-black head and body, and ash-grey wings. Juveniles and winter-plumage birds have mostly white underparts, but are still much darker on the upperside than any of our breeding tern species. The short and notched rather than forked tail is another useful identification feature. When feeding the birds seem tireless, looping, hovering and dipping over the water in pursuit of tiny prey, and rarely dropping into the water.

In the days or even hours after a spring storm, Black Terns often materialise in small groups over lakes and gravel pits, having been blown off course. On the coast they can be seen flying by singly or in small parties during both spring and autumn, with south-east coast being the likeliest regions in which to see them. If they opt to linger and refuel they are likely to head inland to find a lake or other freshwater body rather than to feed at sea, as they are primarily insectivores and prefer to dip feed over still water. However, they can sometimes be seen among the melee of gulls and terns feeding on fish attracted to warm-water outflows, such as that associated with the power station at Dungeness (known to birders as 'The Patch').

Unmistakable in summer plumage, the smoky-toned Black Tern has long been extirpated as a British breeding bird.

Most of the White-winged Black Terns that visit Britain are juveniles, which stray here on their first autumn migration.

The other two *Chlidonias* species both turn up occasionally in the British Isles as vagrants, and both are most likely to be found in the south-east. The **White-winged Black Tern *Chlidonias leucopterus*** looks similar to the Black Tern, but has whiter wings and a shorter bill. The **Whiskered Tern *Chlidonias hybrida*** has a black cap and white cheeks, but smoky-grey underparts, and looks rather like a small, dark, short-tailed Arctic Tern. Both species become much whiter in winter plumage, and identifying a winter-plumage or juvenile *Chlidonias* tern in Britain can be difficult.

All of the marsh terns have a similar ecology, nesting in colonies around shallow lakes and marshland. Often they nest on floating vegetation, and they feed primarily on insects. Like other terns they are migrants and head south for the winter, which is when they may end up in Britain.

The genus *Onychoprion* is a distinctive grouping, sometimes known as the 'brown-backed terns'. These species have dark brown, grey or black uppersides and conspicuous white forehead-patches, and three of the four species found in the world have occurred in the British Isles. The **Sooty Tern *Onychoprion fuscatus*** is a very widespread island breeder, found across the equatorial zone. It is about the size of a Sandwich Tern but is neatly patterned in black and white. Juveniles are also very striking, with entirely dusky blackish-brown plumage, scaled with pale feather fringes. The species has been recorded in both Britain and Ireland, and a few of those birds have associated with tern colonies and lingered – for example, one bird was seen at multiple sites in the County Down and County Dublin area over two months in summer 2005.

The striking Sooty Tern has a large global population, but very rarely wanders as far north as Britain.

The **Bridled Tern *Onychoprion anaethetus*** resembles the Sooty Tern but is smaller, browner and less dapper, with a vague pale collar. Its distribution is similar to the Sooty Tern but a little more northerly. Of the 20 or so records of this species in Britain, several involved tideline corpses, and others were of birds seen among colonies of breeding terns. The third species in the group is the pretty **Aleutian Tern *Onychoprion aleuticus***, a smallish, dark grey-and-white tern that breeds in south-west Alaska, Kamchatka and the Aleutian Islands. The sole individual recorded in Britain, a bird seen among other terns on the Farne Islands in May 1979, was thousands of kilometres away from home and constitutes one of the most mystifying of British rarity records.

Another oddity found on the Farnes, but one that made a lengthy stay and became something of a local celebrity, was 'Elsie' the **Lesser Crested Tern *Thalasseus bengalensis***. Elsie ('L C' for Lesser Crested) paired up with a Sandwich Tern and produced hybrid chicks in 1992–1997, and she or another Lesser Crested Tern was seen at many other sites through successive summers from 1984. Resembling a darkish Sandwich Tern with a yellow bill, the Lesser Crested Tern breeds patchily around the Mediterranean, Red Sea, Arabian Gulf, Indonesia and Australia. The similar and closely related (but orange-billed) **Elegant Tern *Thalasseus elegans*** has been reported on a few occasions in Britain, but has yet to be added to the official British Ornithologists' Union list. This species breeds on the Pacific coast of Mexico and in the southern US; it has a global status of Near Threatened because a large proportion of its small population breeds on just one island (Isla Rasa in the Gulf of California).

Cabot's Tern *Thalasseus acuflavida*, formerly considered a subspecies of the Sandwich Tern, is native to North and South America. There are two subspecies, and the more southerly form is itself sometimes split as a separate species – the **Cayenne Tern *Thalasseus [acuflavida] eurygnatha***. Both Cabot's and Cayenne Terns have occurred in Britain. The sole record of Cabot's Tern is of a first-winter bird found dead well inland in Herefordshire in 1984; luckily this bird was ringed, establishing beyond doubt both its identity and origin. An apparent Cayenne Tern visited a Sandwich Tern colony in Wales in 2006.

The final *Thalasseus* species on the British List is the **Royal Tern *Thalasseus maximus***, a very large species with a shaggy crest and heavy orange bill. It breeds in tropical and subtropical regions, on both American coasts, the Caribbean and east Africa, and birds move both north and south after the breeding season. The first British record was of a bird ringed in the USA and found with other terns in Wales in 1979; there have since been only four other UK records, plus one found dead in Ireland. Even this hefty

The Royal Tern is one of several large, crested, yellow-billed species – correct identification can be challenging.

The largest tern species in the world, the Caspian Tern has a top-heavy look and a huge 'carrot' bill.

tern looks petite alongside the **Caspian Tern** ***Hydroprogne caspia***, the largest member of the family. Nearly the size of a Lesser Black-backed Gull and sporting a huge orange-red bill that is often likened to a carrot, this bird is quite unmistakable. It breeds patchily in eastern Europe and across Asia, North America and Australia. There have been more than 250 records in Britain and Ireland since 1950.

The species most closely related to the giant Caspian Tern actually looks a lot more like a Sandwich Tern, at least at first glance. The **Gull-billed Tern** ***Gelochelidon nilotica*** is a pale, largish tern with a solid black bill that is distinctly shorter and stouter than most terns' bills, hence its name. It breeds mostly in southern Europe and across Central Asia to China, as well as on the Atlantic and Pacific coasts of the USA and northern South America. There have been more than 250 British and Irish records since 1950, and in that year there was a breeding attempt in Essex, which sadly failed when the one chick that hatched died before fledging. Declines of the species' European populations have resulted in fewer birds reaching Britain, and at present it seems unlikely that there will be future breeding attempts.

The final vagrant tern to reach our shores is a North American species, **Forster's Tern** ***Sterna forsteri***. It is closely related to and very similar to the Common Tern, and has occurred about 50 times in the British Isles. The records show a distinct westerly bias to its distribution, and more records in Ireland than elsewhere. It is mainly an inland breeder in its native lands, but the British and Irish records involve birds found on the coast. Most of the records are from between October and March and involve long-staying birds, making this rare visitor perhaps the likeliest tern species to be seen here in the winter.

Gull-billed Terns breed mostly in southern Europe but rarely occur in Britain. They resemble thick-billed Sandwich Terns but develop entirely whitish caps in winter.

Auks

The auks are oddities. They are part of the large order Charadriiformes along with waders, gulls, skuas and terns, but with their pied plumage, chunky shape and upright stance they are much more akin to a certain famous group of flightless seabirds found in Antarctica (among other places). Many non-birders, on seeing a Guillemot for the first time, are convinced that they have found a penguin, and indeed the extinct Great Auk was the first bird species to be known as 'penguin'.

Like their southern-hemisphere lookalikes, auks are squat, short-legged and short-tailed birds, with small and short but strong wings that they use to 'fly' under water. Unlike penguins, auks can fly in the air and are fast fliers, but they are also rather inefficient, having to work very hard to keep airborne, and are much more at ease on or in the water. Their fish-hunting dives are deep and long, aided by their relatively high body density and the extremely effective waterproofing and heat-conserving properties of their sleek, tight plumage. Some can capture multiple fish on a single dive, so that when feeding chicks they can cut down on the number of fishing trips they need to make. Small auks are highly vulnerable to predators and kleptoparasites when flying, so reducing the time spent commuting between sea and nest can be very valuable. Some species nest in burrows as a means to avoid aerial predators, while others are adapted to successfully nest on the most precarious of cliff ledges.

All of our auks are social, nesting in usually large colonies on cliff-faces and clifftops, and they are a very prominent part of most mixed cliffside seabird colonies. Most species have singleton chicks each year, and in some species the young birds leave the nest while still small and unable to fly properly. Each chick must make a heart-stopping leap of faith into the relative safety of the sea, to swim far from land under the care of its father. Because auks spend so much time on the sea's surface, they are very vulnerable to oil spills and other surface pollution.

Worldwide, there are 22 species of auks. The majority are rather small, round birds, and some sport remarkable head and bill ornamentations in the breeding season. Four extant species breed in the British Isles, a fifth occurs regularly as a passage migrant and a few more have been recorded as vagrants, some a staggeringly huge distance from where they should have been. Even the smaller species have long lifespans and great powers of endurance, although in very prolonged rough weather they can suffer high mortality as they struggle to feed successfully.

A Black Guillemot shows off its vivid scarlet feet and mouth.

Guillemot
Uria aalge

The great seabird colonies at Bempton Cliffs, Marwick Head, Fowlsheugh and many other sheer coastal cliffs are home to tens of thousands of Guillemots. These auks are well known for requiring less personal space than other seabirds, living almost pressed up against their next-door neighbours, and for their ability to set up home on the most alarmingly narrow of ledges. At the height of the breeding season whole stretches of cliff-face appear to be made out of Guillemots. When not breeding they roam the seas and may turn up at coastlines anywhere around the British Isles, although any Guillemot that lingers inshore in winter for a long spell may be unwell or oiled.

INTRODUCTION

When seen on land the Guillemot's characteristic upright posture, resting on its ankles, is noticeable, although it also sometimes rests on its belly. The upperparts are dark sooty-brown, and this coloration extends over the whole of the face and as far as the upper breast. The underside is white, neatly demarcated from the dark upperside, and there is a thin white wing-bar. The head and bill have a streamlined shape, with the face smoothly tapering towards the fairly long and strong, dagger-shaped bill. The legs are very short and dark, and set far back on the body. The eyes are dark, and in the 'bridled' morph (which is uncommon but most frequent further north, ranging from 1 per cent of all birds in England to 20 per cent in Shetland) there is a white eye-ring and streak behind the eye. Swimming Guillemots sit fairly low on the water, although not as low as Cormorants, and often look quite hunched and neckless. In flight they look torpedo shaped, and are borne along with very fast flapping of the short, narrow wings. In winter plumage the throat and cheeks are white, but the crown remains dark, with a thin dark line running from the back of the eye downwards across the cheeks. Young Guillemots fresh from the nest are marked like winter adults, but are much smaller and shorter-billed. They are almost invariably seen in the company of an adult Guillemot (their father).

Like the bird on the right of this trio, a minority of Guillemots are 'bridled', with a white spectacle-like marking around each eye.

In spring, thousands of Guillemots rest on the sea by their breeding cliffs.

DISTRIBUTION, POPULATION AND HABITAT

There are Guillemot colonies on suitable cliffs around Scotland, Wales and Ireland. In England there is a long gap in distribution between Yorkshire and Dorset, with a few small colonies in the south-west and some larger ones in north-east England. The biggest single colony is on Handa island off north-west Scotland, which holds more than 100,000 birds, and there are other very large colonies on Berriedale Cliffs in Caithness, Lambay Island off County Dublin, Fowlsheugh in north-east Scotland and Rathlin Island off County Antrim. Overall, Scotland has about 75 per cent of the total population and Ireland about 15 per cent. In England the largest colony (by a sizeable margin) is at Bempton Cliffs in Yorkshire, and the largest in Wales is on Skomer island, Pembrokeshire; England and Wales hold about 6.3 per cent and 3.7 per cent of our total population respectively. There are about 1 million pairs in Britain and Ireland in total.

There are Guillemots on mainland European coasts, although nowhere are they as numerous as in the UK. However, there are very large numbers on Iceland, the Faroes and some Norwegian islands. The distribution extends across Russia and on to both coasts of North America. The total world population is about 18 million birds.

Guillemots nest on sheer cliffs that are inaccessible to mammalian predators, and can use very narrow ledges as well as wider shelves and the tops of stacks. On some islands, where predatory mammals are of no concern, they also nest in caves and on boulder beaches. They spend the post-breeding period on the open sea, and usually feed just offshore, although some are back at their colonies as early as October.

BEHAVIOUR AND DIET

Guillemots on their breeding ledges are not especially active, partly because they are slow and awkward on their feet and partly because they have very little space in which to move. They sit either upright or flat on their stomachs, preen and occasionally argue with their neighbours, which involves pecking and shoving, sometimes to the point that the loser is pushed off the cliff. However, they are generally very tolerant of each other, as they have to be to manage to breed successfully in such close proximity – paired birds preen each other, but may also sometimes even preen their neighbours.

When they set off from the breeding cliff to feed, Guillemots fly quickly down to the sea where they are likely to join many other auks in a 'raft' on the surface. Birds intending to feed assemble at certain points where the nature of the sea and seabed encourages fish to congregate, for example where there are upwellings and abrupt temperature changes due to a sudden change in depth. Most breeding birds forage within 15km of the breeding colony, although at some colonies the average foraging distance can be much further.

In flight, the small size of this bird's wings is very evident – however, it is quite competent in flight once it achieves lift-off.

Taking off from water is hard work, and the birds are highly vulnerable to predators at this time.

Guillemots may rest on the water for some time, using their feet to control their position in choppy seas, and frequently 'standing up' to stretch out and flap their wings. In winter most of their time is passed in this way, alternating long spells of rest with foraging dives. Before diving they frequently submerge their heads, looking for fish nearby.

The Guillemot's actual dive is energetic, with a jump, a stroke of the wings and a strong backwards flick of the feet to propel itself downwards. Under water it 'flies' steeply and quickly downwards, and once at the right level to chase fish it flaps and glides in rapid pursuit, making sharp turns as necessary. Sometimes several birds work together to 'herd' a shoal of fish, each bird snapping up stragglers at the edge of the shoal. The dive depth varies considerably, with a clear distinction between 'short and shallow' and 'long and deep'; the latter can reach 200m and last about four minutes. However, dives to 50m are not unusual.

Guillemots feed on a range of fish types including sandeels, herring and sprats. Different colonies show different dietary breakdowns, indicating that the Guillemot is quite a versatile predator. In some areas they may take a high proportion of crustaceans. Other prey includes squid and polychaete worms.

HOW LOW CAN YOU GO?

The need to be able to fly places limitations on how deep a seabird can dive. Penguins, which have dispensed with flight, are comfortably the deepest divers, as their anatomy is uncompromisingly adapted for swimming and diving. However, the auks are incredibly adept at diving while still hanging on (just) to their ability to fly, enabling them to nest in the safety of tall cliff-faces and other spots that are comfortably out of reach of most predators.

How can we measure the depth to which a bird can dive? A rather crude (and rather sad) source of data comes from recording diving birds found drowned after becoming entangled in gill nets on the seabed. Guillemots have been recovered from nets set as deep as 180m. Depth gauges fitted to birds at one breeding colony recorded dives of 138m, and a study comparing dive duration with swimming speed suggests that the maximum achievable depth is 230m.

Guillemots make use of every available ledge on their cliffs, and have little need for personal space.

BREEDING

Guillemots moult after they have taken to the sea following the breeding season, rendering them briefly flightless. Some may return to their breeding colonies immediately after this, in mid-autumn, especially if they belong to colonies in which prime nesting spots are at a premium. Many others stay at sea for much longer, however, and the colony is not fully occupied again until April. Males are usually back at their nest-sites first, and make loud growling calls to attract the females. Their mates of the previous year seek to rejoin them, and their bonding displays include head dipping and mutual preening of the head-and-neck plumage. However, females that arrive before their mates may be approached by male neighbours and even subjected to forced copulations. Only birds five years old or older are likely to attempt to breed.

In the two weeks before egg laying the female spends most of her time at sea, feeding at a high rate in preparation for forming the single egg that she will lay, while the male remains at the nest-site. In fact these birds do not make an actual nest of any kind, but the female lays her single egg directly onto bare rock. Both birds incubate, taking turns in shifts that can last as long as 38 hours. The incubation period varies from 28 to 37 days, although both laying and hatching tend to be close to synchronous among neighbouring pairs, an adaptation that reduces each chick's chance of predation as predators such as gulls and skuas are 'swamped' with prey for a short time.

Feeding the chick must be done with care to ensure that the youngster does not fall.

Chicks are downy when they hatch, but require brooding by a parent for several days after hatching. The other parent is on foraging duty, bringing fish back to the ledge one at a time, and carefully feeding them to the chick head first. The chick stays close to the back 'wall' of its ledge and holds on to the rock with its large, hooked claws, but by about ten days old it becomes more adventurous and may shuffle along the ledge into the space of the next pair along. This is often tolerated by the neighbours – especially if there are good relationships between them and the chick's parents, perhaps fostered by mutual preening earlier in the season. Sometimes adult Guillemots that have lost their own egg or chick attempt to brood or feed next door's baby. However, there is also sometimes aggression from adults towards unrelated chicks, particularly in years of poor food supply.

At three weeks old the chick is barely half the size of its parents, with a small, stubby bill, but it has its first set of feathers and is waterproof enough to swim safely. It leaves the nesting cliff under

THE SPINNING EGG

Guillemot eggs are noticeably 'top shaped' or pyriform, with the pointed end being very narrow relative to the round end. This has long been supposed to be an adaptation to the species' precarious nesting situation. On a tiny cliff ledge it would not take much of a nudge to send an egg over the edge, but the egg's shape makes it more likely to spin in a circle than roll in a straight line.

Another possible reason for the shape is that it allows the female to comfortably lay a proportionately very large egg (some 11 per cent of her own weight) without compromising her streamlined shape. A large egg results in a large chick that will not take too long to grow to a size where the risk of being taken by a predator is reduced.

The father sticks close to his chick after fledging, and will do so until the young bird is fully grown.

the encouragement of its father, jumping into the air and slowing its fall with spread wings and tail, and splayed feet. Provided it reaches the open water in one piece, it and its father swim out to sea together. This is a spectacle that has caught out many birdwatchers, who report seeing a Little Auk and a Guillemot swimming together – such is the size difference between parent and offspring. The chick can fly after five to seven weeks, but its father continues to feed it for at least another couple of weeks.

MOVEMENTS AND MIGRATION

Guillemots are not true migrants but wander in various directions in winter, with many established adults staying near their colonies, while young birds roam further. A high proportion of English and Scottish birds spend the winter on the North Sea and further east in the Skagerrak and Kattegat straits that separate Denmark from Norway and Sweden respectively. Others head west or south-west towards the Bay of Biscay – there are numerous recoveries of ringed birds from the Atlantic coasts of France and northern Spain. Smaller numbers reach the Baltic Sea and western Mediterranean. There have also been many records of British-ringed birds from Iceland.

Winter-plumaged Guillemots have a mostly white face, and mature adults develop the dark-faced breeding plumage earlier in spring than younger birds do.

THE FUTURE

This auk is increasing in number at present, although there are concerns that climate change could impact on numbers.

Guillemots have shown a continuous increase since the Operation Seafarer survey in 1969–1970. That survey found just over 600,000 individuals in the UK, while the Seabird Colony Register 15 years later found just over a million and Seabird 2000 counted nearly 1.5 million. It is thought that the increase has continued into the 21st century, albeit at a slower rate. They are, nevertheless, in the Amber category of species of conservation concern, because a very high proportion of their population is concentrated at just a few sites. This is a trend that is becoming more marked, along with a general shift northwards. Several small southern colonies have disappeared since the 1970s, while most of the large Scottish and Irish colonies have grown, most notably the one on Rathlin Island, which went from 41,887 birds at the Seabird Colony Register to 95,117 pairs at Seabird 2000. There is a clear limit to the size a colony can reach, as even Guillemots eventually run out of ledge space.

Guillemots are adaptable feeders, which may explain why they have continued to do relatively well when so many other seabird species have not. They have, however, suffered some significant mass mortalities over the last 100 years. They are highly vulnerable to oil spills, and when the *Torrey Canyon* struck rocks and spilled 32 million gallons of crude oil into the Atlantic off south-west England in spring 1967, most of the seabird victims were Guillemots and Razorbills – up to 30,000 of them. In the same region in February 2013, the possibly deliberate discharge of a sticky pollutant identified as polyisobutylene, from an unknown source, killed at least 4,000 auks. Close to our shores, a major oil spill from the *Erika* off Brittany in 1999–2000 killed as many as 120,000 Guillemots.

Guillemots do have some capacity to bounce back from such events. Although each pair only raises one chick a year, the survival rate is high, and if a colony suffers high losses of breeding adults, young birds may breed at an earlier age than normal and thus fill up the spaces. Ongoing problems from food shortages, perhaps linked to climate change, have the potential for more lasting damage, and some Guillemot populations (for example on Bear Island, Norway) have shown declines in correlation with falling fishery catches. However, at present the species in Britain and the wider world appears to be in good shape.

Guillemots are highly vulnerable in the event of oil spills. Clean-up is a long and delicate process.

Razorbill
Alca torda

Guillemots and Razorbills are often spoken of in the same breath, and indeed these two auk species are similar in both appearance and ecology. However, it takes a little more work to find and watch Razorbills. Scan along a cliff-edge packed with Guillemots, and sooner or later you will probably find a lone Razorbill on the edge of a Guillemot cluster, or maybe a small group of Razorbills by themselves, sitting on a wider rocky outcrop than the hair's-breadth ledges favoured by their relatives. They are strikingly handsome birds, dapper in their black-and-white plumage and sporting an impressive, vertically flattened bill. For every 100 pairs of Guillemots breeding in Britain there are only 13 pairs of Razorbills, but they are as much a part of the cliff-side seabird scene as their more abundant relatives. They have also shown a similar general pattern of increase since the late 20th century.

INTRODUCTION

The Razorbill has a very penguin-like aspect – it is stocky, stout and usually seen standing upright with its weight resting on its feet and ankles, or lying on its belly. It has a short, pointed tail and short, narrow and blunt-tipped wings, which are white below. The rest of its plumage is black or blackish-brown above, including the head and throat, and white below, with a small white wing-bar. The head shape tapers towards the bill, which is not especially long but quite deep, flattened on the vertical plane with a hooked tip. The bill is black and marked with a narrow, vertical white stripe halfway down its length; there is also a narrow white stripe running from the base of the upper mandible to just in front of the eye. The eye itself is dark and difficult to make out against the equally dark head plumage. In winter the throat and most of the face become white, leaving just a dark crown and a smudgy dark area around the eye, but there is no dark stripe on the cheek as in the Guillemot. In flight the bird looks front heavy and flies fast, the stubby wings beating in a rapid blur. Young Razorbills are about two-thirds the size of adults when they fledge, with markings similar to those of winter-plumage adults but distinctly smaller bills.

The white facial markings of the Razorbill are noticeable at quite long range, as is its uniquely shaped bill.

This auk is present at most northern seabird colonies but invariably in lower numbers than the Guillemot.

DISTRIBUTION, POPULATION AND HABITAT

The distribution of Razorbills in the British Isles very closely mirrors that of Guillemots. They are widespread along the coasts of Scotland, Ireland and Wales, but in England are restricted to the north-east and south-west. The biggest colonies are on Rathlin Island off County Antrim, Handa island off north-west Scotland and Barra Head in the Western Isles. Bempton Cliffs is the most important site in England, and Skomer island holds the largest colony in Wales. The total population in Great Britain is estimated to be about 126,000 pairs, with another 17,400 in Ireland. This amounts to 17–18 per cent of the world's population.

The Razorbill is distributed across the north Atlantic, reaching Russia in the east and as far south as France (just) and Maine. Iceland holds nearly half of the world's population (about 380,000 pairs); other strongholds are Norway with 20,000 pairs and Canada with 37,800 pairs.

Razorbills mainly nest on cliff-faces, and also on boulder beaches and scree where these are safe from mammalian predators. They tend to use shorter but wider ledges than Guillemots, and often find spots that are very tucked away, in cracks and crannies. They feed in shallow seas over sandy seabeds and where there are upwellings from deeper water; when breeding this is usually 2–15km from the colony, but there is much variation from site to site. Birds seen swimming very close (a few metres) to the shore or inland are likely to be storm-driven, oiled or otherwise in distress.

Head-on, the vertical flattening of the bill is evident. This structure provides a strong bite but keeps the bill relatively lightweight.

BEHAVIOUR AND DIET

During the breeding season Razorbills that are not busy incubating or provisioning their chicks spend most of their time resting either on ledges or on the sea below, often in the company of (and outnumbered by) Guillemots in both cases. Their journeys between cliff and sea are typically rapid and direct, as flight is so energetically expensive for them. Razorbills at some colonies forage very close inshore, while at others they move 20km away or more. Birds at Skomer, for example, sometimes go as far as 45km from 'home'. Most feeding dives take them no more than 30m down, but dives of up to 140m have been recorded.

Like other auks Razorbills dive from the surface when hunting, and swim under water with strong strokes of their flipper-like wings. They may submerge the head several times before diving, to check the positions of fish near the surface, or may dive immediately upon landing on the water, presumably having spotted fish while flying down to the sea. They are more likely to forage alone than are Guillemots, but do also act cooperatively (perhaps with Guillemots) to drive a panicking 'ball' of fish towards the surface, then attack it from below.

Pairs conduct tender bonding rituals, but this seabird will not spurn the chance of a casual liaison with a neighbour.

The diet is mainly composed of sandeels, sprats, herring, rocklings, gobies and some crustaceans. The proportions of each in the diet vary according to location. In Scotland Razorbills (and other seabirds) have taken an increased proportion of Snake Pipefish during the breeding season in recent years, reflecting a dramatic population increase in this fish species since around 2003.

BREEDING

Razorbills mainly return to their breeding colonies in March. They show strong year-on-year fidelity to their colony, nest-site and mate. However, female Razorbills do seek copulations with males other than their mates, by visiting particular areas outside the main colony where several non-breeding Razorbills congregate. A paired male does not make any real attempt to prevent his mate from participating in these liaisons, and indeed may seek to mate with other females himself, but initiates frequent copulations with his regular mate (about three per day in the month before egg laying) to try to raise the odds that the single egg she lays will be fathered by him. The female spends nearly all her time feeding at sea over the few days before laying her egg, and most females lay their eggs in May.

The nest-site is bare rock, on a broad ledge (usually with some kind of sheltering overhang), in a crevice or among boulders. The egg is proportionately large and has a rather elongated and pointed shape, although less markedly so than that of the Guillemot. It is whitish with many irregular dark

A & E FOR AUKS

Many auks die at sea from various causes, and are discovered washed up on the shore. Sometimes auks that are still alive are discovered on beaches, and they are likely to be doomed if not rescued and taken to a specialised wildlife hospital, perhaps via a vet. Some such birds will have got into difficulties because they could not find enough food, but others may have been victims of pollution spills, especially oil. When petroleum-based oils get into the plumage they interfere with a bird's natural waterproofing, so that it cannot swim or dive easily and is at high risk of chilling.

The safest way to catch an auk, or any seabird for that matter, is to drop a coat or towel over it, wrap it up gently, then pick up the bundle, holding it at arm's length and keeping the bird's head covered as it may well strike at you. In 2011 a man lost an eye while attempting to rescue an injured Gannet, and even small species like Puffins have a powerful bite. Ideally, the bird should then be transferred to a secure cardboard box before it is taken to a vet or directly to a wildlife hospital with suitable facilities for caring for seabirds.

Treating starving auks is straightforward enough, but cleaning oil from the plumage is a difficult task. Repeated washing with a gentle detergent, then thorough rinsing, is needed. Medicine may also be required to counter the effects of oil that a bird has ingested as it has tried to clean itself. Many oiled birds need to be tube fed on highly calorific and easily digested foods in addition to having free access to fish, and have to be kept warm as the oiling and washing process seriously compromises their ability to thermoregulate. They have access to a warm 'therapy pool' where they can preen and swim, and where their carers can check how well their natural waterproofing is recovering, as well as their general alertness and well-being. Once all is satisfactory they are moved to cool outdoor pools for further monitoring before release.

The whole process is lengthy and expensive, and many rehabilitated birds undoubtedly do not survive long after release. However, most are ringed, and ringing recoveries have shown that some do make a full recovery and survive for many years after treatment.

Nest-sites for Razorbills are typically more sheltered and on wider ledges than in the case of Guillemots.

speckles and squiggles. Both adults incubate it for about 36 days. If an accident or predation befalls the egg the female may lay a replacement, but this is usually smaller than the first, and the chances of the parents successfully hatching a replacement egg and rearing the chick are reduced.

The chick is fed and tended by both parents, which feed it on increasingly large fish as it grows, delivering them one at a time. The chick hatches with a coat of down and open eyes, and rapidly develops a serviceable set of body feathers that keep it warm and dry when it takes to the sea at the tender age of about 18 days. In the company of its father it jumps, scrambles or climbs to the sea, and the two then head out to deeper water where they will be safer from gulls, skuas and other predators that would have no difficulty taking the small chick. It is about two-thirds of adult size at this point, but swims comfortably and is fed by its father.

Razorbills begin breeding at the age of three, four or five. They are very long-lived seabirds. The oldest ringed bird in Britain was ringed as a chick on Bardsey Island in 1962, and was aged almost 42 when found again, alive and well, back at Bardsey. Like most seabirds that have long potential lifespans and low productivity, investing heavily in a single chick per year, Razorbills are unlikely to breed every year but instead save their efforts for years when they are in the best physical condition.

The chick has a more conventional bill shape when it is very young, but the adult bill form quickly develops.

A CHANGE THAT IS HARD TO SWALLOW

Pipefish are remarkable-looking creatures, with very slender but extremely long bodies (up to 60cm in the case of female Snake Pipefish). They are related to seahorses and like them have delicate elongated faces, but they lack the seahorses' spines and angular shape. Another trait they have in common with seahorses is that after fertilisation the males nurture the developing young inside a brood pouch, 'giving birth' to them in due course.

Snake Pipefish numbers in the north-east Atlantic increased at a staggeringly rapid rate in the first few years of the 21st century. It is not known what triggered the increase, although possible causes include climate change (warmer seas being more conducive to successful breeding). Their prevalence among items offered by seabirds such as Razorbills to their chicks has increased accordingly, especially in areas affected by shortages of sandeels, a staple breeding season prey for many seabird species. However, the pipefish are difficult items for chicks to manage, being very long and having relatively tough bodies, and in colonies there have been many finds of dead seabird chicks that have choked to death on pipefish meals. The combination of fewer sandeels and more pipefish is a dangerous one for many species of seabird.

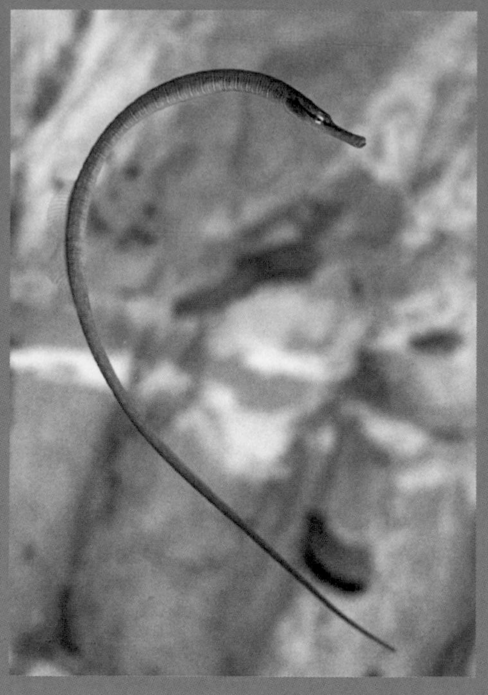

Pipefish are tough-bodied, offering little in the way of nutrition compared to sandeels and clupeid fish.

MOVEMENTS AND MIGRATION

All Razorbills leave their colony in July and head for the sea; each male that bred successfully will be in the company of its chick. The adults moult through late summer, while the chicks are still developing their flight feathers, so both adults and chicks are unable to fly until September. Young birds disperse further than adults, although neither can be said to be truly migratory – they may head either north or south, or remain fairly close to their breeding grounds.

Razorbills in winter plumage have white faces, which gradually become blackish through early spring.

Built for diving rather than flying, this auk can nevertheless cover long distances.

Ringing recoveries of chicks ringed in the British Isles have shown that the birds can disperse as widely as Greenland, the Western Sahara, Italy and the far north of Norway. The majority of recoveries are from the Atlantic coast from southern Norway through Denmark, Germany, the Netherlands, Belgium, France, Spain and North Africa.

THE FUTURE

The three main seabird surveys of the late 20th century found a strong increase in Razorbill numbers in Britain and Ireland, with an increase of 16 per cent between Operation Seafarer (1969–1970) and the Seabird Colony Register (1985–1988), and a further increase of 21 per cent between the Seabird Colony Register and Seabird 2000. The rate of increase has slowed since then with a rise of 3 per cent in 2000–2012. Broken down by country, in Scotland numbers have increased steadily; in England they increased sharply until the 1980s and since then have been stable; in Wales the reverse is true with stability until the mid-1980s, then a strong increase; in Ireland numbers fell between Operation Seafarer and the Seabird Colony Register, but have since risen strongly.

Part of the increase may simply be down to improved counting methods. Razorbills are difficult to count, as they often breed in hidden spots or are mixed in with Guillemots. However, the increase in Razorbills has been in step with that of the easier-to-count Guillemot, which suggests that it is genuine, as the two species have a quite similar ecology. Razorbills can feed at a wide range of depths in the sea, so have a certain adaptability compared with surface feeders, which has allowed them to cope with the shortage of sandeels, for example, better than Kittiwakes and terns have done. On the other hand, deep diving places them at risk of entanglement in gill nets, which can be a source of significant mortality.

The risk of mass mortality from oil-spill events is ever present, and Razorbills suffered high losses in the *Torrey Canyon* disaster, for example. There have also been a number of significant 'wrecks' of Razorbills, probably because of bad weather and associated difficulties with foraging. In September 2007 a mass wreck of seabirds on the coasts of south-east Norway was composed mainly of Razorbills, and other events took place in the North Sea in 1995 and 2013. The Isle of May Razorbills showed a sharp fall in adult survival rates after the 1995 event, from a mean of 90.7 per cent to just 73 per cent. However, counts on the island during the 1996 breeding season revealed higher numbers of Razorbills than in any year since 1963, despite the losses of the year before. Clearly there is much to learn about the population dynamics of this species, but for now it appears to be coping well with the many challenges of life for a seabird in the 21st century.

Black Guillemot *Cepphus grylle*

This attractive auk is unlike our other three breeding species in that it is not a long-distance wanderer, but remains close to its 'home' colony at all times. It also tends to establish colonies of just a few pairs at sites away from the main seabird cities, although there are often a few on suitable cliffs as well. One of the so-called 'true guillemots', it is a rounder and less penguin-like bird than the other British auks, with unmistakable (although very different) plumage patterns in the breeding season and in winter. It is the least numerous of our auks and probably the most difficult of them to accurately survey, because of the scattered nature of its breeding sites.

INTRODUCTION

The Black Guillemot in breeding plumage has velvety blackish-brown plumage with a large white oval on the wing-coverts – this allows easy identification even if the bird is on the water and its dark underside cannot be seen. Its shape is less elongated than that of the Guillemot, and its bill is shorter. The monochrome plumage is relieved by its bright red legs and feet, and it also shows a scarlet gape when the bill is opened. In winter it is very different, with almost white plumage over most of the body, speckled and scaled with black, and with the eyes and bill looking very black. In this plumage it could be mistaken for a seaduck or grebe at first glance. It still shows black flight feathers, and a clean white oval on the wing. Birds moulting from breeding to winter plumage show increasingly extensive white speckling, beginning on the face. On the sea the bird looks buoyant and its short tail is often cocked, although its wings may droop into the water. In flight it looks rather duck-like and not quite as stubby winged as the Guillemot or Razorbill, and shows white linings to the undersides of the wings.

The Black Guillemot is shorter-bodied and longer-winged than the larger auks, and consequently more agile in the air.

This species readily uses artificial structures for resting on, and sometimes even for nesting.

DISTRIBUTION, POPULATION AND HABITAT

This species is distributed along the north and west coasts of Scotland, on most of the Irish coast and on the Isle of Man. It is scarce away from these areas – only a handful of pairs breed in England (at St Bees Head in Cumbria), and in Wales there are a few breeding sites in the north of the country (mainly on Anglesey). Scotland holds about 70 per cent of the breeding population and Ireland nearly all of the rest, with a total population of just over 42,000 individuals.

Black Guillemots breed on Arctic coasts around northern Europe, Russia and across to Alaska, the Atlantic coast of Canada and the northern USA, on Greenland and on Iceland. Canada holds the largest population with about 71,000 pairs; there are some 55,000 pairs in Russia, 30,000 in Iceland and 20,000 in Norway. The total world population is 400,000–700,000 individuals. The most northerly birds move south for the winter, while others (including British birds) are very sedentary.

Black Guillemots tend to nest among rocks on small islands and low-lying coastlines, rather than on tall and sheer cliff-faces, and their nest-sites are often well hidden and (where there are no mammalian predators) even on the ground. At several sites they nest on artificial structures such as piers and jetties. They habitually feed closer inshore than do our other auks.

TIZZIE THE TYSTIE

Along with the Great Skua, aka the Bonxie, the Black Guillemot is a species known as often by an affectionate nickname as it is by its full official English name. The Shetland name for the Great Skua, Tystie, is widely used among birders and others familiar with the species. Shetlanders have their own names for many seabirds, but most of these have not really caught on beyond the islands. Among them are Dunter for the Eider, Tirrick for the Arctic Tern (an accurate rendition of that species' call), and Tammie Norie for the Puffin.

Lillian Beckwith's book *Beautiful Just!* is the sixth book she wrote about her life as an amateur crofter on the tiny island of Soay, just off the Isle of Skye. In this book she tells of encountering a swimming Black Guillemot fledgling while out rowing on the sea on a hot summer's day. After rescuing the chick from the attentions of a hungry 'black-backed gull', she found that it would not stop following her boat – the perhaps inevitable consequence was that the young bird became her pet and lived in the croft with her despite being free to come and go if it wished. It was named Tystie, or Tizzie for short. She wrote with great charm and humour about the bird's quirky personality and intelligence – it was quick to learn how to manipulate its human carers in order to obtain extra helpings of fish.

TRUE GUILLEMOTS

Although we tend to think of our Guillemot as the 'default guillemot' because it is the most familiar bird to us that bears the name, in fact it and the related Brünnich's Guillemot are known as murres in North America and some other areas. The three species in the genus *Cepphus* are officially the 'true' guillemots.

Besides the Black Guillemot there are the Pigeon Guillemot *Cepphus columba* and the Spectacled Guillemot *C. carbo*. The Pigeon Guillemot is very similar to the Black Guillemot but a little larger, with a steeper forehead (giving it a 'pigeon-like' appearance) and a dark bar across its white wing-patch. The Spectacled Guillemot is all-dark apart from whitish 'spectacle' markings around its eyes. Both species occur on the opposite side of the world, the Pigeon Guillemot on the Pacific coasts of Siberia and North America, the Spectacled Guillemot around the north-west Pacific. The Pigeon Guillemot has occurred in Norway and could theoretically stray to British waters, although it would take a skilled and attentive birdwatcher to identify it because it is so similar to the Black Guillemot.

BEHAVIOUR AND DIET

This auk is less gregarious than our other species and is usually seen in pairs or small parties, and quite often alone when feeding. It stands in a more horizontal posture than the Guillemot, and habitually sits back on its ankles or on its belly. With proportionately broader wings and a smaller overall body size, it looks more agile and confident in the air than the larger auks. It spends much time resting on the water. Black Guillemots remain close to their breeding grounds throughout the winter and may be seen resting on shoreline rocks or structures projecting into the sea at any time of the year.

The species is a pursuit diver and propels itself with its wings when swimming under water. It tends to hunt in shallower water than other auks – typically shallower than 35m – and seeks prey near the seabed. It very rarely descends below 45m and usually forages within 5km of its nesting site. Occasionally, several birds work with each other to 'herd' fish, diving together along a straight line or semi-circle. Birds looking for prey dip their heads under water before diving.

The Black Guillemot's diet contains a higher proportion of non-fish prey than the diets of other auks – it catches crabs and marine worms on the seabed, and also takes molluscs and sea scorpions. However, fish still form the bulk of its diet in most circumstances, with favoured types including sandeels, butterfish and various kinds of flatfish.

This auk feeds on a wide range of marine animals, rather than being almost exclusively fish-eating like its relatives.

Birds display at the start of the breeding season on prominent outcrops, rushing about with much calling.

BREEDING

Breeding activity begins in late March, with adults gathering in small groups on land to display to each other before establishing pair bonds. The display involves standing or running in a 'tall' posture with wings half open and repeatedly giving a thin, whistling call. This display exhibits the red feet and mouth, and the white wing marking, to their best advantage. As pairs form the courtship display also includes slow circling in the same posture, and a pair also performs a circling display on the water. It is during these periods of display that surveyors count Black Guillemots – once nesting many birds are too well concealed for accurate counts from either land or the sea.

Year-on-year mate fidelity is high – a Danish study found that it ranged from 64 to 79 per cent. Loyalty to previously used nest-sites is also high, but the chances of a bird opting to return to its natal colony to breed for the first time are rather low.

Nest-sites are usually sheltered, often in gaps between boulders on beaches, or in caves. Anywhere that offers shelter and concealment, as well as safety from high tides, is potentially suitable. Pairs may nest in quite close proximity if the lie of the land is conducive – each pair only defends an area of a little under a metre radius around the nest. In the nest crevice the female lays her clutch, sometimes on bare rock and sometimes on a bed of small pebbles. The clutch is usually of one or two eggs, occasionally three. The gap between laying is three days, and incubation begins with the arrival of the second egg. This places the youngest chick in a brood of three at an obvious disadvantage and its survival is unlikely.

Broods of two rather than one are quite common in this auk.

Incubation lasts for about 30 days. Should the eggs be lost within the first week or so the female will probably lay a replacement clutch. The parents both develop brood patches and share incubation duties, taking shifts of a few hours each, and the off-duty bird often 'stands guard' close to the entrance of the nest-site. The chicks are downy when they hatch, with open eyes and enough strength in their legs to crawl about. As is typical in auks they are well developed, although they do not leave the nest at quite such an early age as Razorbills and Guillemots.

The chicks are brought whole fish to eat, which the parents carry crosswise in their bills one at a time. It has been noted that the adults show a kind of 'handedness', in that each bird always carries its fish with the head pointing in the same direction. For very small chicks, dead fish are delivered, but after a few days the fish are brought alive and wriggling. The chicks squeak and jump about until they are fed, and until they are about three weeks old they wait for the next meal at almost the exact spot where the eggs were laid. Past this age they begin to tentatively explore the rest of the nest cavity and to spend time stretching their wings and preening.

The adults feed the chicks for about 35–40 days, and by this time they are close to full size and ready to live independently, although they are not yet able to fly. The adults may lure them out into the open with a dangling fish. Once out they make their way to the sea by climbing and jumping, and swim out to a suitable depth for feeding.

MOVEMENTS AND MIGRATION

Black Guillemots are the most sedentary of British breeding auks, and are unlikely to be encountered away from nesting areas at any time of year. The arrival of a Black Guillemot offshore in any county away from where the species breeds is usually notable enough to attract a crowd of local birdwatchers. The majority of ringing recoveries have been of birds found or resighted close to where they were originally ringed. The longest distance travelled by a ringed bird was 877km; the bird in question was ringed as a chick on Fair Isle in 1953 and shot in the Blackwater Estuary in Essex in the autumn of the same year. This is an exceptional record as the bird not only travelled a great distance, but also moved well away from any sites where the species breeds.

THE FUTURE

Black Guillemots were not counted adequately during the Operation Seafarer survey (1969–1970), as the count took place in June when many of the birds were out of view at their nests. For the Seabird Colony Register (1985–1988) and Seabird 2000, counts were made of birds near breeding colonies between 26 March and 15 May, and only on calm days; many counts were from boats. In Scotland the numbers recorded by Seabird 2000 were almost identical to those in the Seabird Colony Register, increasing from 37,172 to 37,505 individuals. Numbers on the Isle of Man and in Northern Ireland doubled between the two surveys, from 303 to 309 birds, and 533 to 1,174 birds respectively. Pairs in England and Wales remained very low (in England 14 birds were counted in the Seabird Colony

Like other auks, the Black Guillemot often lets its wings droop open into the water when swimming.

Although less familiar than our other auks, this is a delightful and charming bird with a currently stable British population.

Register and seven in Seabird 2000, while in Wales the figures were 26 for the Seabird Colony Register and 28 for Seabird 2000). Unfortunately no data were collected for the Republic of Ireland in the Seabird Colony Register – Seabird 2000 counted 3,367 birds. No data is yet available for the period post-2002.

Although the population has stayed broadly stable since 1969, there have been some losses of colonies and some new ones have been established. There is a general suggestion of a southwards push – most colonies on the Isle of Man are either increasing or have been fairly recently established, and the same is true of Wales and Northern Ireland. There has been a fall in numbers on Orkney (15 per cent between the Seabird Colony Register and Seabird 2000), although not on Shetland where numbers rose by 31 per cent over the same period.

The distribution and feeding behaviour of Black Guillemots is quite different from that of other auks. They are perhaps more at risk from oil spills as they feed closer inshore at all times of the year. The *Braer* disaster off Shetland took place in January 1993 and killed an estimated 1,300 Black Guillemots, far more than all other auk species put together. However, Black Guillemots have higher productivity than other auks and can thus recover more quickly from losses like this.

Some of the colony losses in western Scotland are thought to be the result of American Mink predation. Mink can cross more than a kilometre of open sea so even island populations are at risk. Otters are also potential predators, but the small size of cavities used by Black Guillemots often keeps them out; unfortunately, mink are considerably smaller and can enter any opening that a Black Guillemot can. Eradication of mink on some islands has allowed lost or declining Black Guillemot colonies to recover.

Another cause of mortality is entanglement in fishing nets, but this accounts for only a small proportion of deaths. Losses associated with food availability do not seem to be a factor either. The diverse diet of Black Guillemots is probably key to their current and future survivability.

Puffin
Fratercula arctica

With its clown face, preposterous multicoloured bill and general air of pompous solemnity, the Puffin is one of our most beloved and familiar birds, and certainly our most recognisable seabird, even though most people have never seen one in the flesh. Puffins breed patchily around the coastline of much of northern England, Wales, Scotland and Ireland, with a few colonies holding tens of thousands of birds, and may be seen offshore at other times of the year. The smallest of our breeding auks, they keep themselves safe while nesting by using burrows, but are still vulnerable to attack by gulls and skuas. They have undergone a population increase over recent decades, especially in England.

INTRODUCTION

A Puffin in full breeding plumage is difficult to confuse with any other bird, even when seen from some distance. Black above with a white face and belly but black breast-band, its appearance is dominated by the deep, vertically compressed bill with its pattern of red-and-blue stripes. The dark, triangular marking above, behind and in front of the eye gives it a characteristic anxious expression, coupled with the downturned yellow flanges on the gape sides. The feet and legs are bright orange-red, and the bird's overall shape is compact, with short wings and tail and a large-headed, top-heavy look. In winter the Puffin becomes a little less distinctive, as the colourful bill plates are shed, leaving a smaller and duller bill, and also the face becomes dusky grey. Other auks, however, become whiter faced in winter. Juvenile Puffins are also grey faced, with even smaller and drabber bills. In flight the top-heavy appearance is even more noticeable, and the rather narrow wings (proportionately longer than those of Razorbills and Guillemots) beat very rapidly. On the water the Puffin has a buoyant and short-bodied look.

Puffins are rather ungainly in flight but make good use of updraughts to control their approach to land.

This is an auk of clifftops rather than cliff-faces.

DISTRIBUTION, POPULATION AND HABITAT

Puffins are most numerous in north-east and northern Scotland, especially St Kilda, Orkney and Shetland. There are also colonies scattered along the east coast of Scotland and down to Yorkshire, then there is a long gap around the east and south-east coasts. There are a few colonies in the south-west, and more on islands off Pembrokeshire and North Wales. The species is rather thinly distributed in north-west Scotland and occurs patchily around Ireland, mostly on the west coast. There are an estimated 600,000 pairs altogether, 20,000 of them in Ireland and nearly 500,000 in Scotland.

Puffin colonies can hold hundreds of pairs or more, and their burrows can cause significant erosion.

The Puffin's official English name is Atlantic Puffin, to distinguish it from the other two puffin species in the genus *Fratercula*, and the name accurately describes its distribution. It breeds throughout the north Atlantic, from Newfoundland and Greenland, across via Iceland to mainland Europe across Norway, with a few birds reaching Russia and a few ranging south as far as the Canary Islands. The world population is estimated to be 6.6 million pairs, so that the British Isles hold a fairly significant 10 per cent or so of the global population.

Puffins need soft ground in which to dig their burrows (or existing Rabbit burrows that they can take over), so they usually colonise the earthy tops of cliffs or stacks. They also sometimes nest in crevices on boulder beaches. They usually fish within 5km of their colonies when breeding, but sometimes go much further (up to at least 40km), and make use of tidal fronts where the flow gradient brings suitable prey closer to the surface.

ALSO IN THE PUFFIN SERIES

The two other *Fratercula* species are unmistakably puffins, with the trademark large and colourful bills. They are, if anything, even more outlandish in appearance than 'our' Puffin, the Horned Puffin (*F. corniculata*), having an even larger, mostly yellow bill and strong eye markings that give it a very severe expression, and the Tufted Puffin (*F. cirrhata*) sporting outrageous back-swept blond eyebrows. Both occur in the North Pacific, but the Tufted Puffin has strayed to British waters on one occasion (see page 234).

When not actively feeding or incubating, Puffins often prefer to rest on the sea rather than on land.

BEHAVIOUR AND DIET

In typical auk style, Puffins spend much of their time resting on the sea, often in the company of other Puffins, or (in the breeding season) standing near their nesting burrows. In winter they are less gregarious, and often feed alone.

On the breeding grounds Puffins often stand very upright, with their weight on just their feet rather than the feet and ankles as is the case with larger auks. They can walk, run and jump more comfortably than the other auks, although they do have a rather comical waddling gait; they also rest on their bellies at times. They are a little more capable in the air than larger auks, although they still need to flap hard and fast to stay aloft when not assisted by updraughts. When coming in to land on a clifftop they can almost hang on updraughts to steady themselves before landing, using both wings and splayed feet to hold their position.

Puffins are surface divers and 'fly' under water with strong but shallow beats of their wings, interspersed with glides, and no or very little movement of their feet. They can catch multiple fish in a single feeding dive, and most of the time under water is spent at 20–30m under the surface, swimming quickly backwards and forwards in pursuit of fish. Deeper dives of just over 60m have been recorded.

While sandeels are the best-known prey favoured by Puffins, their prevalence in the diet varies greatly according to location and time of year. They also take Saithe, whiting, Capelin, herring, rocklings, and very young cod and Haddock. Non-fish prey is uncommon but may include small crustaceans.

Taking off from the sea is a challenge even for this relatively lightweight auk.

The celebrated Puffin bill is a useful tool for digging and for pulling up grass to line the nest, as well as for catching fish.

BREEDING

Puffins return to their clifftop breeding grounds in March or April. Those that have bred successfully before seek to reoccupy the same burrow year on year, and bond with the same partner. Courtship involves a head-waggling, bill-bumping display, accompanied by harsh growling calls. Working together to renovate the burrow also helps to cement the bond between the couple. It is a common sight in spring to see one member of a pair standing guard outside a burrow and ignoring the spray of dust that is being kicked into its face by its partner, who is working away unseen in the tunnel. For early arrivers in the north of the species' range, the burrow may need to be cleared of snow before it can be used.

For new pairs and those that have lost their burrows (or seek a new one because nesting failed the year before), a new burrow must be excavated. Puffins are skilled diggers, vigorously using both bill and sharp-clawed feet to excavate a tunnel about 90cm deep. However, they also use suitable existing hollows if they can find them, including those dug by Rabbits and Manx Shearwaters, and some nest

HOW MANY SANDEELS?

Puffins are well known for being able to carry a large quantity of sandeels or similar small fish in their bills at once, all neatly packed in sideways-on – more than 60 on occasion. The anatomy of the bill is such that more fish can be caught when the bill is already half full, making for very efficient deliveries to the chick. The riskiest time is the flight from the sea back to the nest-site, when skuas and gulls may harry Puffins and try to relieve them of their catch. When pursued a Puffin may let go of some of its cargo, in the hope that the pursuer will be diverted and the Puffin freed to return to its burrow with at least a partial meal for its chick. However, it might be best off relinquishing all of its fish quickly if it is targeted by a Great Skua or Great Black-backed Gull, as these large hunters are more than capable of catching and killing the Puffin itself. Any contact with the predator could be disastrous for the Puffin, because if its fast-flapping flight is interrupted it will probably plummet earthwards, and may be unable to recover flapping speed in time to avoid hitting the cliff-side.

in hollows between boulders. The key is to make or use a hole that is small and deep enough to be inaccessible to avian predators such as crows and skuas. Puffin burrows are still vulnerable to attack from small mammalian predators like rats, and the most successful Puffin colonies are on rat-free islands and stacks.

A Puffin ensconced in its burrow is vulnerable but will attack intruders with a powerful peck, as many bird ringers have discovered to their cost.

By mid-May all is ready and the female lays her single egg. The parents share incubation duties, the off-duty bird spending its time feeding at sea or resting near the burrow entrance. Activity among these birds tends to be synchronous, so at times an active colony can seem deserted as half the birds are out of sight incubating in their burrows and the other half are far out at sea. Sometimes the non-incubators wheel around the colony en masse, a dramatic display that is thought to have some role in discouraging avian predators.

The egg (which is whitish, as is common with hole-nesting birds, rather than heavily marked like the other auks' eggs) takes 36–43 days to hatch. The chick is down covered but rather helpless, and remains in its burrow until the day it fledges and makes the long drop to the open sea, aged 38–44 days.

Unlike Razorbills and Guillemots, adult Puffins provide no further care to their chicks after fledging, although they feed them very generously up until that time. When it is ready to leave the nest the young Puffin, a dark-faced, small-billed version of the adult, leaves the burrow under cover of darkness (the best time to avoid predatory birds) and walks to the cliff-edge. Then it half falls and half flies down to the sea. Instinct takes over, and the youngster swims with ease. Mastering the arts of diving and fishing may take longer, but the chick has ample fat stores to see it through a lean time while it develops its skills. It is, of course, particularly vulnerable to predators in its first weeks of independent life, and many chicks fall prey to gulls and skuas, but once it reaches deep offshore water it is safer. It spends its first winter offshore and, while it may visit the colony in subsequent summers (although usually just joining off-duty breeding birds on the sea rather than coming ashore), in most cases it is not ready to breed until it is five years old. Individual Puffins can live into their thirties, and most of the birds that survive well into adulthood take at least one or two years off from breeding in their lifetimes.

A 'puffling' close to fledging is much drabber and smaller-billed than its parents.

MOVEMENTS AND MIGRATION

Where Puffins go in the winter is not fully understood. Ringed British birds have been found washed up on the coastlines of France, Spain, Italy, Sweden, Denmark, North Africa, the Cape Verde Islands, Iceland, Greenland and eastern Canada. Analysis of ringing data suggests that birds breeding in eastern Britain winter primarily in the North Sea, but other British Puffins apparently range across the Atlantic and into the Mediterranean. Presumably the birds concentrate wherever fishing conditions are best, but no particular consistent 'hotspots' have been identified.

In a bid to learn more researchers fitted geolocating trackers to 50 Puffins from the Isle of May, north-east Scotland, in 2009. Their data revealed that these birds were moving between the North Sea and the western Atlantic, more so than the ringing data predicted. This may indicate a new behavioural shift, in response to changes in fish stocks. Another 25 Puffins tracked from a southern Irish colony in 2012 were found to make a full Atlantic crossing, possibly to exploit abundant stocks of Capelin, one of their preferred fish.

No seabird is as instantly recognisable, and the Puffin often headlines stories about seabird conservation issues.

THE FUTURE

Most key Puffin sites are under protection, and many colonies, especially those in northern England, have seen marked increases since the late 20th century. Overall, there was a rise of 15 per cent between the Operation Seafarer survey in 1969–1970 and the Seabird Colony Register in 1985–1988, and a further rise of 19 per cent between the Seabird Colony Register and Seabird 2000. England in particular saw a very strong increase, of 676 per cent between Operation Seafarer and Seabird 2000. Much of this increase took place on the Farne Islands and Coquet Island in Northumberland, but Bempton Cliffs in Yorkshire also saw a strong increase. Numbers in Scotland have increased by more modest amounts, with increases on Shetland counterbalanced by declines on Orkney, and the huge St Kilda colony remaining fairly stable. In Ireland numbers have been down by about 18 per cent since Operation Seafarer; in Wales (the country in the UK with the fewest breeding Puffins) they have increased by about 167 per cent over the same period.

Predation by rats and minks can be a problem for Puffins – the arrival of rats on Ailsa Craig led to the near extinction of Puffins there, but they were quick to recolonise after rat eradication in 1990–1991. Oiling incidents and drowning in gill nets accounts for small numbers; the most serious recent incident was the sinking of the *Prestige* oil tanker off north-west Spain in November 2002. An estimated 2,200 Puffins were killed as a result, presumed to be British breeding birds, but even this had no noticeable impact at colonies.

Nesting in burrows provides a certain level of security from predation, but burrows can be flooded out with disastrous consequences in bad summer weather, or at some sites by unusually high tides. Ground that is used by many Puffin pairs can also become destabilised to the point that burrows collapse in on themselves, and further tunnelling on the eroded ground becomes impossible.

Puffins in North Sea areas have felt the effects of sandeel declines, and in some years have shown very poor breeding success, in common with other birds like Kittiwakes and terns. As the bad years have been interspersed with better years the overall impact has not been severe, but this and other food-supply issues in the breeding season are likely to be the most serious threats to Puffins in the future.

Predation by Great Skuas accounts for losses of many Puffin chicks in northern Scotland.

Other auks

The **Great Auk *Pinguinus impennis*** has the dubious honour of being the only bird on the British List that is known for certain to be extinct in the wild, although a couple of curlew species are likely to have joined it on the 'permanently missing' list since the mid-20th century. What were probably the last two individuals, and last known breeding pair, were killed at their nest-site on an island off Iceland on 3 July 1844. Although many factors came together to bring about the bird's catastrophic decline, its own rarity was the final nail in its coffin, as collectors frantically set about obtaining eggs and specimens when it became obvious that the species was on the brink of extinction.

Great Auks resembled over-sized Razorbills, with a similar bill shape and plumage pattern, but they had a prominent circular white marking on the forehead. In winter plumage further white areas developed on the face. The wings were tiny; this species was flightless, and pursued a lifestyle similar to that of the penguins, spending most of its time on the water and breeding in large colonies on rocky beaches. It would have been a wing-propelled diver, and without the need for flight was probably a deep diver, rivalling the large penguin species.

Great Auks are known to have bred on the coasts of Britain and Ireland, as well as south to France and northern Spain, further east to Norway, and on the other side of the Atlantic in Canada and the northern US states. Little is known about their ecology beyond that which can be inferred from their anatomy and similarity to Razorbills.

It is clear from the prevalence of Great Auks' remains at ancient sites that they were of considerable importance to local human populations as far back as 100,000 years ago. Hunting gradually intensified from the 8th century onwards, with the birds being exploited for meat, eggs and down feathers, and even being used as fishing bait and fuel for fires – explorers reported that their oily bodies burned quite satisfactorily. By the 16th century it was obvious that the population had declined catastrophically, and

Tiny, round and small-billed, the Little Auk looks rather like a newly fledged Guillemot or Razorbill chick.

Lloyd, C., Tasker, M. L. & Partridge, K. 1991. *The Status of Seabirds in Britain and Ireland* (Poyser Monographs). Poyser, London.

Machovsky-Capuska, G. E., Howland, H. C., Vaughn, R. L., Würsig, B., Raubenheimer, D., Hauber, M. E. & Katzir, G. 2012. Visual accommodation and active pursuit of prey underwater in a plunge diving bird: the Australasian Gannet. *Proceedings of the Royal Society: Biological Sciences* 279: 4118–4125.

Malling Olsen, K. & Larsson, H. 2004. *Gulls of Europe, Asia and North America* (Helm Identification Guides). Christopher Helm, London.

Matthews., G. V. T. 1954. Some aspects of incubation in the Manx Shearwater *Procellaria puffinus*, with particular reference to chilling resistance in the embryo. *Ibis* 96: 432–440.

Meek, E. R., Bolton, M., Fox, D. & Remp, J. 2011. Breeding skuas in Orkney: population change driven by both food supply and predation. *Seabird* 24: 1–10.

Meyburg, B.-U. & Meyburg, C. 2009. Wanderung mit Rucksack: Satellitentelemetrie bei Vögeln [Complete English translation: Travels with a backpack: satellite tracking of birds]. *Der Falke* 56: 256–263.

Monaghan, P., Uttley, J. D., Burns, M. D., Thaine, C. & Blackwood, J. 1989. The relationship between food supply, reproductive effort and breeding success in Arctic Terns *Sterna paradisaea*. *Journal of Animal Ecology* 58 (1): 261–274.

Nelson, B. 1978. *The Gannet*. Poyser, Berkhamsted.

Newell, D. 2008. Letter: recent records of southern skuas in Britain. *British Birds* 101 (8): 439–441.

Ollason, J. C. & Dunnet, G. M. 1978. Age, experience and other factors affecting the breeding success of the Fulmar, *Fulmarus glacialis*, in Orkney. *Journal of Animal Ecology* 47 (3): 961–976.

Öst, M. & Jaatinen, K. 2013. Relative importance of social status and physiological need in determining leadership in a social forager. *PLOS ONE (Public Library of Science)* 8 (5): e64778/1–7.

Perrins, C. M., Wood, M. J., Garroway, C. J., Boyle, D., Oakes, N., Revera, R., Collins, P. & Taylor, C. A. 2012. Whole-island census of the Manx Shearwaters *Puffinus puffinus* breeding on Skomer Island in 2011. *Seabird* 25: 1–13.

Piatt, J. F. & Nettleship, D. N. 1985. Diving depths of four alcids. *The Auk* 102 (2): 293–297.

Potts, G. R. 1966. *Studies on a marked population of the shag (*Phalacrocorax aristotelis*), with special reference to the breeding biology of birds of known age*. Doctoral thesis, Durham University.

Quinn, J. L., Still, L., Carrier, M. C., Kirby, J. S. & Lambdon, P. 1996. Scaup *Aythya marila* numbers and the Cockle *Cardium edule* fishery on the Solway Firth: are they related? *Wildfowl* 47: 187–194.

Ratcliffe, N., Schmitt, S., Mayo, A., Tratalos, J. & Drewitt, A. 2008. Colony habitat selection by Little Terns *Sternula albifrons* in East Anglia: implications for coastal management. *Seabird* 21: 55–63.

Reading and Basingstoke Ringing. 2010. *Berkshire Black-headed Gull ringing project 2010 report*. Berkshire.

Ross, B. P. & Furness, R. W. 2013. Minimising the impact of eider ducks on mussel farming. Ornithology Group, Institute of Biomedical and Life Sciences, Graham Kerr Building, University of Glasgow, Glasgow G12 8QQ.

Schreiber, E. A. & Burger, J. 2001. *Biology of Marine Birds*. CRC Press, Florida.

Seltmann, M. W., Jaatinen, K., Steele, B. B. & Öst, M. 2013. Boldness and stress responsiveness as drivers of nest-site selection in a ground-nesting bird. *Ethology* 119: 1–13.

Sjölander, S. 1978. Reproductive Behaviour of the Black-throated Diver *Gavia arctica*. *Ornis Scandinavica* 9 (1): 51–65.

Smith, A. J. M. 1975. Studies of breeding Sandwich Terns. *British Birds* 68: 142–156.

Status of Great and Arctic Skuas on Handa island: www.handaskuas.org/conservation

Survey of breeding Little Terns in Wells Harbour, Norfolk: www.wellsharbour.co.uk/na545.htm

Taverner, J. H. 1965. Observations on breeding Sandwich and Common Terns. *British Birds* 58: 5–9.

Tierney, N., Lusby, J. & Lauder, A. 2011. A preliminary assessment of the potential impacts of Cormorant *Phalacrocorax carbo* predation on salmonids in four selected river systems. *Report Commissioned by Inland Fisheries Ireland*.

Townsend, C. W. 1909. The use of the wings and feet by diving birds. *The Auk* 26 (3): 234–248.

Tree, A. J. 2011. Origins, occurrence and movements of Sandwich Tern *Thalasseus sandvicensis* in southern Africa. *Marine Ornithology* 39: 173–181.

UK policy on Cormorant control at fisheries:

http://archive.defra.gov.uk/wildlife-pets/wildlife/management/policy/licensing-cormorants.htm

Votier, S. C., Bearhop, S., Newell, R. G., Orr, K., Furness, R. W. & Kennedy, M. 2003. The first record of Brown Skua *Catharacta antarctica* in Europe. *Ibis* 146 (1): 95–102.

Wakefield, E. D., Bodey, T. W., Bearhop, S., Blackburn, J., Colhoun, K., Davies, R., Dwyer, R. G., Green, J. A., Grémillet, D., Jackson, A. L., Jessopp, M. J., Kane, A., Langston, R. H. W., Lescroël, A., Murray, S., Le Nuz, M., Patrick, S. C., Péron, C., Soanes, L. M., Wanless, S., Votier, S. C., Hamer, K. C. 2013. Space partitioning without territoriality in gannets. *Science* 5 (341): 68–70.

Watson, M. J., Spendelow, J. A. & Hatch, J. J. 2012. Post-fledging brood and care division in the Roseate Tern (*Sterna dougallii*). *Journal of Ethology* 30: 29–34.

Acknowledgements

I would like to thank Julie Bailey at Bloomsbury for commissioning this book, allowing me to fulfil a lifelong ambition of writing at length about some of the bird species I have loved most since childhood. Thank you to all the talented photographers whose work appears in these pages, in particular to David Tipling for supplying the lion's share of the wonderful photographs that bring this book to life. Thanks to Krystyna Mayer for her work copy-editing the text, and to Julie Dando for creating the book's beautiful design and for handling all issues that arose in the layout process with her trademark speed, skill and helpfulness. Thanks to RSPB seabird expert Euan Dunn, Mark Boyd, Derek Niemann and the RSPB's Wildlife Enquiries team for checking over the book plans and page proofs, and to Sara Hulse for reading and correcting the final pages.

This book would not have been possible without the efforts of hundreds of researchers and thousands of volunteers, who monitor, study and protect Britain's wonderful seabird colonies. My thanks go to them for all their work to safeguard what is surely one of our most precious and important natural treasures. Finally, I would like to thank my parents for having the great good sense to raise me in a seaside town, thus instilling in me a lifelong passion for the sea and for all sea life.

Photographic credits

Bloomsbury Publishing would like to thank the following for providing photographs and for permission to reproduce copyright material. While every effort has been made to trace and acknowledge all copyright holders, we would like to apologise for any errors or omissions and invite readers to inform us so that corrections can be made in any future editions of the book.

Photographs, including both cover photographs, were supplied by and remain the copyright of David Tipling, except as listed below:

21 Neil Bowman/FLPA; 22T Roger Wilmshurst/FLPA; 23 Mike Lane/FLPA; 24T Roger Tidman/FLPA; 24B Erni/Shutterstock; 26T Erni/Shutterstock; 26B David Hosking/FLPA; 28 Erni/Shutterstock; 29 Feathercollector/Shutterstock; 31 Paul Reeves Photography/Shutterstock; 38 V. Belov/Shutterstock; 39T Robin Chittenden/FLPA; 40 Erni/Shutterstock; 41 Erni/Shutterstock; 43B Erni/Shutterstock; 45T Elizabeth Hoffmann/Shutterstock; 46 Erni/Shutterstock; 47 Mark Medcalf/Shutterstock; 48 Marianne Taylor; 50B Erni/Shutterstock; 58 JPS/Shutterstock; 59 RSPB/NOF/Terje Lislevand; 60 Erni/Shutterstock; 62T John Hawkins/FLPA; 65 David Hosking/FLPA; 66 Robin Chittenden/FLPA; 69 Robin Chittenden/FLPA; 69T John Hawkins/FLPA; 71 Steve Young/FLPA; 72 Steve Young/FLPA; 73 John Hawkins/FLPA; 74 S Jonasson/FLPA; 76 John Holmes/SS; 77 Robin Chittenden/FLPA; 78 David Koloechter/Shutterstock; 79 Krisgillam/Shutterstock; 80 BMJ/Shutterstock; 90 Marianne Taylor; 92T Marianne Taylor; 92B Pam Garland/Shutterstock; 93T Erni/Shutterstock; 93B Marianne Taylor; 98 Marianne Taylor; 102 Anatoliy Lukich/Shutterstock; 103T Robert J Richter/Shutterstock; 103B Bikeriderlondon/Shutterstock; 107 Sue Robinson/Shutterstock; 111 Wolfgang Kruck/Shutterstock; 112 Andrew Parkinson/FLPA; 115 Jack Chapman/FLPA; 118T Andrew Parkinson/FLPA; 122 Andrew Parkinson/FLPA; 124 BMJ/Shutterstock; 125 Sebastian Kennerknecht/Minden Pictures/FLPA; 130 John Carnemolla/Shutterstock; 132T Wolfgang Kruck/Shutterstock; 132B Marianne Taylor; 134 Marianne Taylor; 141T Marianne Taylor; 145T Dave Montreuil/Shutterstock; 145B Kateryna Larina/Shutterstock; 146 Neil Bowman/FLPA; 147 Marianne Taylor; 148T Bildagentur Zoonar GmbH/Shutterstock; 148B MirrorOnTao/Shutterstock; 152T Andrew Astbury/Shutterstock; 154B Marianne Taylor; 155T Chris Pole/Shutterstock; 156 Fabio Lotti/Shutterstock; 159 Alexander Erdbeer/Shutterstock; 160T Marianne Taylor; 160B Marianne Taylor; 161B Bill Coster/FLPA; 162 Marianne Taylor; 166 Marianne Taylor; 167B Marianne Taylor; 168B Steve Young/FLPA; 169T Steve Young/FLPA; 169B Txanbelin/Shutterstock; 170B Marianne Taylor; 171B Stubblefield Photography/Shutterstock; 174 Georgios Alexandris/Shutterstock; 176T /FLPA; 176B Panu Ruangjan/Shutterstock; 177T Panu Ruangjan/Shutterstock; 177B Yossi Eshbol/FLPA; 178 Mike Lane/FLPA; 179T Yoshi Eshbol/FLPA; 179B BlurAZ/Shutterstock; 181 Marianne Taylor; 183B Michael Gore/FLPA; 186 Noah Strycker/Shutterstock; 187 Paul Sawer/FLPA; 188 Steve Young/FLPA; 189 Paul Sawer/FLPA; 192T Marianne Taylor; 192B Marianne Taylor; 193 Marianne Taylor; 195 Menno Schaefer/Shutterstock; 196 Dave Head/Shutterstock; 198T Arto Hakola/Shutterstock; 198B Arto Hakola/Shutterstock; 203T Marianne Taylor; 203B Ashley Whitworth/Shutterstock; 213B Corepics VOF/Shutterstock; 214 Menno Schaefer/Shutterstock; 216 Menno Schaefer/Shutterstock; 217T JBernspang/Shutterstock; 218T D P Wilson/FLPA; 218B Marianne Taylor; 222 Harri Taavetti/FLPA; 223B Imagebroker/FLPA; 224 Marianne Taylor; 228T Richard Costin/FLPA; 233 Rebecca Nason; 234T Robert L Kothenbeutel/Shutterstock; 234B Gary Thoburn.